symposia on
theoretical
physics
and mathematics

9

Contributors to this volume:

P. W. Coulter
P. P. Divakaran
D. A. Dubin
W. H. J. Fuchs
R. H. Good, Jr.
A. Kumar
A. M. Lee
S. Okubo
G. Rajasekaran
R. Ramachandran
A. Ramakrishnan
L. A. Rubel
G. L. Shaw
S. Srinivasan
R. Vasudevan

symposia on theoretical physics and mathematics

Lectures presented at the
1968 Sixth Anniversary Symposium
of the Institute
of Mathematical Sciences
Madras, India

Edited by
ALLADI RAMAKRISHNAN
Director of the Institute

 PLENUM PRESS • NEW YORK • 1969

ISBN-13: 978-1-4684-7675-0   e-ISBN-13: 978-1-4684-7673-6
DOI: 10.1007/ 978-1-4684-7673-6
Library of Congress Catalog Card Number 65-21184

Introduction

This volume represents the proceedings of the Sixth Anniversary MATSCIENCE Symposium on Theoretical Physics held in January 1968 as well as the Seminar in Analysis held earlier, in December 1967. A new feature of this volume is that it includes also contributions dealing with applications of mathematics to domains other than theoretical physics. Accordingly, the volume is divided into three parts—Part I deals with theoretical physics, Part II with applications of mathematical methods, and Part III with pure mathematics.

The volume begins with a contribution from Okubo who proposed a new scheme to explain the CP puzzle by invoking the intermediate vector bosons. Gordon Shaw from Irvine dealt with the crucial importance of the effects of CDD poles in partial wave dispersion relations in dynamical calculation of resonances. Applications of current algebra and quark models were considered in the papers of Divakaran, Ramachandran, and Rajasekharan. Dubin presented a rigorous formulation of the Heisenberg ferromagnet.

In view of the renewed interest in theories of particles with arbitrary spin, this volume contains the lectures of Good on a new approach to the problem of electromagnetic interaction of such particles. While the theory of wave equations for particles of spin greater than $1/2$ is one direction in which the Dirac equation is capable of extension, it was also felt appropriate to point out a new direction in which the Dirac algebra could be extended from a Clifford algebraic point of view. In three papers, the contributions on this subject from MATSCIENCE were presented establishing the relationship of higher dimensional extensions of the Dirac matrices with spinor theory and Clifford algebra.

Lee's lecture on operational research is the special feature of this volume and provides a link between mathematics and industry. Srinivasan presented a survey of some aspects of stochastic integration and differential equations while Vasudevan discussed the role of stochastic point processes in understanding coherence properties of laser beams.

In the section on mathematics we are particularly grateful to Professor

Rubel for his valuable contribution "Vector Spaces of Analytic Function" which dealt with the interaction between the theory of analytic functions and functional analysis. In a second article, Rubel discussed the uniform distributions and densities on locally compact Abelian groups. Fuchs in his contribution discussed the application of Nevanlinna theory to gap series. This is an interesting sequel to that of Hayman who has given a very good introduction to Nevanlinna theory in Vol. VI of these proceedings.

Contents

THEORETICAL PHYSICS

APPLICATIONS OF MATHEMATICAL METHODS

PURE MATHEMATICS

Contents of Other Volumes

VOLUME 1

VOLUME 2

VOLUME 3

VOLUME 4

VOLUME 5

VOLUME 6

VOLUME 7

VOLUME 8

VOLUME 10

(Partial Listing)

Intermediate Vector Mesons and *CP*-Violations

S. Okubo

UNIVERSITY OF ROCHESTER
Rochester, New York

The weak interactions have provided many surprises, the first of which was the violations of the parity (P) and the charge conjugation (C) invariance and, more recently, the *CP*-violation. The so-called $V-A$ theory proposed by many authors is a very elegant theory which states that the P and C violations occur maximally, so to speak. However, recently discovered violation of the CP (or time-reversal invariance by *TCP* theorem) is more puzzling, since its effect appears to be very small by a factor of 10^{-3} in comparison with the *CP*-conserving one. To explain this, some authors have suggested to give up the $V-A$ theory by inserting a tiny relative phase-factor between V and A parts. But this is not only artificial but also spoils the beauty of the original $V-A$ theory. Another possibility is simply to assume the existence of a weaker *CP*-violating interactions, which is again a rather an artificial way without really explaining the reason for its smallness. Here, I would like to propose a new scheme where the *CP*-violation is maximal as well as maximal violations of separate P and C invariances. At a first glance, this may appear to be quite impossible since experimental *CP*-violation is smaller by a factor 10^{-3} as has been mentioned already. But fortunately, we could make a consistent theory of this kind as will be explained below.

Our fundamental postulates are as follows:

1. There exist intermediate vector mesons (hereafter, W-bosons) and they have a strong trilinear interaction among themselves. However,

1

it has been assumed that they have only weak and electromagnetic interactions with all other particles.

2. The weak interaction Hamiltonian $\mathscr{H}_W$ is linearly proportional to w-meson field, i.e., it is a Yukawa-type interaction. Further we require that $\mathscr{H}_W$ has $CP = -1$.

To be more specific, let us schematically write

$$H_w = ig(j_\mu + l_\mu)W_\mu + \text{h.c.}$$

where j_μ and l_μ are hadronic and leptonic currents, respectively. Then, the ordinary CP-conserving weak interactions come from the second-order effect g^2, while the CP-violation results in the third-order g^3 which is nonzero because of our first postulate. Note that in this scheme, the first-order effect proportional to g is absent for ordinary processes involving no W-boson. Thus, our theory is very similar to that of multipole expansion in atomic or nuclear spectroscopy. Indeed, if we use their terminology, then, the CP-conserving and effects roughly correspond to E_1 and M_1 transition, respectively, while the dominant E_0 transition is forbidden by gauge invariance.

Before going into details, it may be worthwhile to point out that the existence of self-couplings among W-bosons is really welcome since it may account for the large mass of these bosons if they exist. Again this fact is based upon ordinary philosophy that the physically observed masses (or at least a major part) of elementary particles result from self-energy corrections, so that the hadron masses are heavier than that of the electron (although the relatively large mass of muon is still a bit of mystery). However, the existence of self-coupling as has been postulated in (1) leaves still too much arbitrariness in its exact form. Thus, in this note we tentatively assume further that this self-coupling is a result of Yang–Mill-type character of the W-mesons. To be more definite, let us assume that we have an octet of hermitian w-bosons fields, $W_\mu^{(\alpha)}(x)$, $(\alpha = 1, \ldots, 8)$ and define an octet of self-conjugate hadronic currents $j_\mu^{(\alpha)}$, $(\alpha = 1, \ldots, 8)$ analogously. In the quark model, $j_\mu^{(\alpha)}(x)$ may be given by

$$j_\mu^{(\alpha)}(x) = \frac{i}{2}\, \bar{q}(x)\gamma_\mu \lambda_\alpha (1+\gamma_5)q(x) \tag{1}$$

although in what follows, its explicit form is really not essential.

Note that $j_\mu^{(\alpha)}(x)$ has the structure of the exact V–A form and that

$$[j_\mu^{(\alpha)}(x)]^+ = \begin{cases} j_\mu^{(\alpha)}(x) & \mu \neq 4 \\ -j_4^{(\alpha)}(x) & \mu = 4 \end{cases} \tag{2a}$$

Similarly,

$$[W_\mu^{(\alpha)}(x)]^+ = \begin{cases} W_\mu^{(\alpha)}(x) & \mu \neq 4 \\ -W_4^{(\alpha)}(x) & \mu = 4 \end{cases} \tag{2b}$$

Now, for a later convenience, let us introduce a traceless tensor

$$(W_\mu)_b^a(x) \qquad (a, b = 1, 2, 3)$$

by

$$(W_\mu)_b^a(x) = \frac{1}{\sqrt{2}} \sum_{\alpha=1}^{8} (\lambda_\alpha)_b^a W_\mu^{(\alpha)}(x) \tag{3}$$

so that the hermiticity requirement is now given as

$$[W_\mu(x)]_b^{a+} = \pm W_\mu(x)_a^b$$

according to $\mu \neq 4$ or $\mu = 4$. Then, the inverse formula is

$$W_\mu^{(\alpha)}(x) = \frac{1}{\sqrt{2}} \sum_{a,b=1}^{3} (\lambda_\alpha)_a^b (W_\mu)_b^a \tag{4}$$

In equations (3) and (4), λ_α ($\alpha = 1, \ldots, 8$) are the 3×3 matrices introduced by Gell-Mann.

Similarly, one can define a traceless tensor $(j_\mu)_b^a(x)$ with $a, b = 1, 2, 3$ by

$$(j_\mu)_b^a(x) = \frac{1}{\sqrt{2}} \sum_{\alpha=1}^{8} (\lambda_\alpha)_b^a j_\mu^{(\alpha)}(x)$$

$$j_\mu^{(\alpha)}(x) = \frac{1}{\sqrt{2}} \sum_{a,b=1}^{3} (\lambda_\alpha)_b^a (j_\mu)_a^b(x) \tag{5}$$

Then, $(j_\mu)_b^a(x)$ is easily seen to be given by

$$(j_\mu)_b^a(x) = \frac{i}{\sqrt{2}} \bar{q}_a(x)\gamma_\mu(1 + \gamma_5)q_b$$

$$-[\text{trace with respect to } SU(3)] \tag{6}$$

in the quark model.

The reason why we use the tensor notation is two-fold; first that it gives a more compact way of writing and second that the charge-conjugation is much simpler in its form.

Now, we mentioned that $W_\mu^\alpha(x)$ is Yang–Mill type field. By this we mean that we have the following free Lagrangean $\mathcal{L}_0$:

$$\mathcal{L}_0 = -\frac{1}{4} F_{\mu\nu}^\alpha(x) F_{\mu\nu}^\alpha(x) - \frac{1}{2} m_0^2 W_\mu^\alpha(x) W_\mu^\alpha(x)$$

$$F_{\mu\nu}^\alpha(x) = \frac{\partial}{\partial x_\mu} W_\nu^\alpha(x) - \frac{\partial}{\partial x_\nu} W_\mu^\alpha(x) + f_0 \cdot f_{\alpha\beta\gamma} W_\mu^\alpha(x) W_\mu^\beta(x) W_\nu^\gamma(x) \tag{7}$$

so that $\mathscr{L}_0$ contains a trilinear self-coupling among W-bosons as is required in the postulate (1), where we adopted the ordinary convention that a repeated index implies an automatic summation over relevant spaces.

In terms of the tensor notation, we may regard $(W_\mu)_b^a[(F_{\mu\nu})_b^a]$ as 3×3 matrix $W_\mu(F_{\mu\nu})$ for $SU(3)$-index $a, b = 1, 2, 3$ by

$$(W_\mu)_{ba} \equiv (W_\mu)_b^a \tag{8}$$

then equations (3) and (7) are expressible as matrix equations:

$$W_\mu(x) = \frac{1}{\sqrt{2}} \sum_{\alpha=1}^{8} W_\mu^\alpha(x)\lambda_\alpha \tag{9a}$$

$$F_{\mu\nu}(x) = \frac{\partial}{\partial x_\mu} W_\nu(x) - \frac{\partial}{\partial x_\nu} W_\mu(x) - \frac{i}{\sqrt{2}} f_0 \left[W_\mu(x), W_\nu(x) \right] \tag{9b}$$

$$L_0 = -\frac{1}{4} tr[F_{\mu\nu}(x)F_{\mu\nu}(x)] - \frac{1}{2} m_0^2 tr[W_\mu(x)W_\mu(x)] \tag{9c}$$

Note that the equation (9b) for $F_{\mu\nu}$ contains a commutator $[W_\mu, W_\nu]$ as 3×3 matrices. Now if we define the charge-conjugation operation for W-mesons by

$$C: \qquad W_\mu(x) \rightarrow -W_\mu^T(x) \tag{10a}$$

or in terms of components:

$$C: \qquad (W_\mu)_b^a(x) \rightarrow -(W_\mu)_a^b(x) \tag{10b}$$

then $\mathscr{L}_0$ is invariant under the charge conjugation. We state that a charge-conjugation operator for $W_\mu^\alpha(x)$ is a bit more complicated in contrast to the tensor notation. Similarly, if we define the parity operation P by

$$P: \qquad W_\mu(x) \rightarrow \begin{cases} -W_\mu(-\mathbf{x}) & \mu \neq 4 \\ +W_0(\mathbf{x}) & \mu = 4 \end{cases} \tag{11}$$

then $\mathscr{L}_0$ is again invariant under P. Thus, $\mathscr{L}_0$ is separately invariant under both C and P operations as the strong interactions should be. Note also that our assignment for C and P quantum numbers is exactly the same as those for the ordinary vector octet.

On the contrary, the hadronic current $(j_\mu)_b^a(x)$ does not have a simple transformation property under C and P operations but under the product CP, it transforms as

$$CP: \qquad j_{\mu b}^a(x) \longleftrightarrow \begin{cases} j_{\mu a}^b(-\mathbf{x}) & \mu \neq 4 \\ -j_{4a}^b(\mathbf{x}) & \mu = 4 \end{cases} \tag{12}$$

so that we shall call hereafter that $j_\mu(x)$ has $CP = +1$. Then, $W_\mu(x)$ has $CP = +1$ parity also.

Before going into detail, we can state that $\mathscr{L}_0$ is invariant under $SU(3)$ transformation among W-mesons. Since this $SU(3)$ has nothing to do with the ordinary broken $SU(3)$ symmetry of hadrons, we call the former the weak $SU(3)$ and the latter the strong $SU(3)$ group. Then, our assumption for the Lagrangean $\mathscr{L}_0$ given by equation (7) or (9) implies that the weak $SU(3)$ is exact in contrast to the broken strong SU (3) symmetry.

Under these preliminary preparations, we now proceed to the consideration of our second postulate (2) by assuming that the weak interaction Lagrangean $\mathscr{L}_1$ has the following form:

$$\mathscr{L}_1 = ig \{W_1^2[aj_2^1 + bj_3^1 + \alpha l] - W_2^1[aj_1^2 + bj_1^3 + \alpha \bar{l}]$$
$$+ W_1^3[cj_2^1 + dj_3^1 + \beta l] - W_3^1[cj_1^2 + dj_1^3 + \beta \bar{l}]$$
$$+ W_2^3[ej_3^2 + fj_1^1] - W_3^2[ej_2^3 + fj_1^1]\} \tag{13}$$

where we omitted the Lorentz index for simplicity and the leptonic currents l_λ and $\bar{l}_\lambda$ are defined by

$$l_\lambda(x) = i\{\bar{\nu}_e(x)\gamma_\lambda(1 + \gamma_5)e(x) + \bar{\nu}_\mu(x)\gamma_\lambda(1 + \gamma_5)\mu(x)\}$$
$$\bar{l}_\lambda(x) = i\{\bar{e}(x)\gamma_\lambda(1 + \gamma_5)\nu_e(x) + \bar{\mu}(x)\gamma_\lambda(1 + \gamma_5)\nu_\mu(x)\} \tag{14}$$

Moreover in equation (13), all constants $g, a, b, c, d, e, f, \alpha$, and β have been assumed to be real. Then, it is easy to find that $\mathscr{L}_1$ has $CP = -1$. This is the reason why we inserted the imaginary factor i in front of equation (13).

Since both W_b^a and j_b^a are self conjugate [see equation (2)] and since C and P properties of W_b^a are determined solely by $\mathscr{L}_0$ alone, the $CP = -1$ character of $\mathscr{L}_1$ is nontrivial and it cannot be transformed away by suitable definitions of W_b^a and j_b^a or of C and P operations for W_b^a and j_b^a. Further, it may be remarked that we included a neutral current proportional to j_3^2 and j_1^1 in equation (13), whose presence is essential for our theory. At any rate, our theory implies the maximum CP-violation.

Now, first we notice that the first-order effect in the order g does not give any weak interaction since $\langle W_\mu^\alpha(x)\rangle_0 = 0$ identically due to the Lorentz covariance of theory where the expectation value is to be understood with respect to the vacuum state of the w-boson, with the Lagrangean $\mathscr{L}_0$. Actually, one can prove the stronger result that any combined effect of both electromagnetic and weak interactions will not give rise to any observable result in this order g since

$$g \cdot e^n \langle j_w(x_1) \cdots j_w(x_n) W_2^3(x) \rangle_0 = 0 \qquad (15)$$

for arbitraray n where $j_w(x)$ represents a part of electromagnetic current involving only W-bosons and where we omitted the Lorentz indices again. Equation (15) can be proved either by the conservation of the weak strangeness quantum number or by the weak U-spin conservation since $j_w(x)$ will have $U = 0$, $Y = 0$ while W_2^3 has $U = 1$, $Y = 1$. The validity of equation (15) is essential for our scheme since it ensures the vanishing of electric dipole moment of the neutron in the order g.

Next, let us consider the second-order effect. Then neglecting $SU(3)$ index as well as other multiplicative constants for simplicity, we have to evaluate the following quantity:

$$\mathcal{H}_2 = g^2 (j_\mu(x) + l_\mu(x))(j_\nu(y) + \bar{l}_\nu(y)) \langle W_\mu(x) W_\nu(y) \rangle \qquad (16)$$

This reproduce the ordinary CP-conserving weak interactions for leptonic, semileptonic and nonleptonic processes. We may remark also that if we choose $ab + cd + ef = 0$ then the resulting CP-conserving nonleptonic weak interaction with $\Delta S = \pm 1$ will have an octet transformation property with respect to the ordinary strong $SU(3)$ symmetry. However, the part with $\Delta S = 0$ cannot possess the same property. Moreover, if it is so wished, one can reproduce the Cabbibo theory by setting $\alpha = \beta = 1$, $a = c = \cos \theta$, $b = d = \sin \theta$ for instance.

For the order g^3, similarly, we have to compute the matrix element of the form

$$\mathcal{H}_3 = g^3 (j_\mu^\alpha + l_\mu)(j_\nu^\beta + \bar{l}_\nu) j_\lambda^\gamma \langle W_\mu^\alpha(x) W_\nu^\beta(y) W_\lambda^\gamma(z) \rangle_0 \qquad (17)$$

Since the w-bosons contain a trilinear self-coupling, $\langle W_\mu W_\nu W_\lambda \rangle \neq 0$ in general and we obtain $CP = -1$ interaction in the order g^3. More explicitly, because of the exact validity of the weak $SU(3)$ symmetry together with the fact that W has $C = -1$, we must have

$$\langle W_\mu^\alpha W_\nu^\beta W_\lambda^\gamma \rangle_0 \propto f \cdot f_{\alpha\beta\gamma} \qquad (18)$$

where f represents the renormalized coupling constant corresponding to given in equation (9). Note that comparing equation (16) and (17) we find that the effective CP-conserving and CP-violating interactions $\mathcal{H}_2$ and $\mathcal{H}_3$ are roughly of the order of

$$\mathcal{H}_2 \sim \frac{g^2}{m^2}$$

$$\mathcal{H}_3 \sim \frac{g^3}{m^3} \frac{f}{m^3} \qquad (19)$$

where m is the physical mass of the intermediate vector bosons. Hence if in the unit $m_N = 1$ (m_N being the nucleon mass) we have $f/m^3 \sim$ 0(1), and $g/m \sim 10^{-3}$, then our theory could reproduce a rough order of magnitudes for the *CP*-violation required by the experiment. It is interesting that in our theory we could have *CP*-violations for both semi-leptonic and nonleptonic weak interactions. A more careful inspection shows the following characteristic of the *CP*-violations in the order g^3:

1. For nonleptonic decays, the product of three hadronic currents $j_\mu^\alpha j_\nu^\beta j_\lambda^\gamma$ must have a completely antisymmetric permutation symmetry with respect to any exchange of the strong $SU(3)$ indices. This is due to equation (18) together with the fact that the time-ordered product is symmetric with exchanges of coordinates. Hence, we have to form a completely antisymmetric combination out of tensor product $8 \otimes 8 \otimes 8$. The answer is relatively easy and we find

$$(8 \otimes 8 \otimes 8)_{\text{antisym}} = 1 \oplus 8 \oplus 27 \oplus 10 \oplus \overline{10} \qquad (20)$$

with respect to the $SU(3)$-decomposition. Therefore, the *CP*-violating nonleptonic interactions are sum of $8, 10, \overline{10}, 27$ with respect to the strong $SU(3)$. Particularly, this indicates that

(a) $\Delta S = \pm 2$ interactions could exist but its presence may be eliminated if we choose $b \cdot c = 0$.

(b) $\Delta S = \pm 1$ terms consist in general of both $\Delta I = \frac{1}{2}, \frac{3}{2}$ part but not $\Delta I = \frac{5}{2}$. The absence of the last is welcome since then one can reconcile the experimentally observed near-equality of the decay rates for $K_\pm \to \pi_\pm \pi_\pm \pi_\mp$. We note that by *TCP* theorem, we can prove the equality of $\Gamma(K_+ \to \pi_+ \pi_+ \pi_-)$ and $\Gamma(K_- \to \pi_- \pi_- \pi_+)$ if there is no $\Delta I = \frac{5}{2}$ and if the linear energy expansion for the decay matrix elements is good. We notice also that the present experimental knowledge of the *CP*-violations for $K_2^0 \to \pi_+ \pi_-$ and $\pi_0 \pi_0$ is not precise enough to allow us to say anything more.

(c) $\Delta S = 0$ terms have only $\Delta I = 1$ and $\Delta I = 0$ but no $\Delta I = 2$ or 3. Experimentally, this is very difficult to test. Also, because of the precence of many unknown parameters $a, b, c, d, e,$ and f, the *CP*-violating $\Delta S = 0$ part can be made smaller compared to that for $\Delta S = \pm 1$. Then, this could account for the smallness of the experimental electric dipole moment of the neutron. Also, from the study of the ordinary β-decays, the upper-limit for the time-reversal violation is known to be of the order of 1% which is still to large compared to our effect of $g \sim 10^{-3}$.

(d) We can have two-types of semileptonic CP-violations of the form $j_\mu j_\nu l_\lambda$ and $j_\mu l_\nu \bar{l}_\lambda$. Hence we expect to have CP-violations in semileptonic decays such as $\Lambda \rightarrow pe\bar{\nu}_e$ or K_{e_4} decays. But its experimental detection up to the accuracy of 10^{-3} will be very difficult. The occurance of the $j_\mu l_\nu \bar{l}_\lambda$ term is a bit more interesting. A more careful inspection shows that it will lead to a decay $K_1^0 \rightarrow \mu\bar{\mu}$ or $e\bar{e}$(but not K_2^0 $\rightarrow \mu\bar{\mu}$ or $e\bar{e}$) contracting neutrino fields. Hence, we expect to have a decay $K_1^0 \rightarrow \mu\bar{\mu}$ (or ee) smaller by a factor of 10^{-6} compared to the normal decay mode $K_1^0 \rightarrow 2\pi$. Note that we can expect to have also the CP-conserving decay for $K_1^0 \rightarrow \mu\bar{\mu}$ (or $e\bar{e}$) in the order $g^2 e^4$ which would give a smaller contribution by a factor as much as 100 compared with our CP-violating effect. Hence, its experimental check would be a test for the validity of our theory. Similarly, CP-violating $K_+ \rightarrow \pi + \mu\bar{\mu}$ or $\pi + e\bar{e}$ could happen in the order g^3. But we expect for this decay to have CP-conserving decay mode of the order $g^2 e^2$ which would be now 100 times larger than the former. Also, we could have $\nu_\mu + p \rightarrow \mu + \bar{e} + \nu_e + p$ or $K_1^0 \rightarrow e + \bar{\nu}_e + \bar{\mu} + \nu_\mu$ in order g^3, but these would be much harder to detect experimentally.

In summary, our theory seems to be consistent with the present experimental facts as far as a rough qualitative estimate is concerned. The present theory also has a certain affinity with a theory recently proposed by Nishijima. However, it may be emphasized that his theory deals exclusively with the nonleptonic process alone, and it cannot treat all leptonic, semileptonic, and nonleptonic interactions in a unified fashion as has been done in this note.

CDD Effects in Partial Wave Dispersion Relations and the Dynamical Calculation of Resonances*

PHILIP W. COULTER and GORDON L. SHAW

UNIVERSITY OF CALIFORNIA
Irvine, California

1. INTRODUCTION

An almost overwhelming number of resonances are now known to exist. In the pion–nucleon system alone, recent phase shift analyses have revealed a total of 19 resonances at masses up to 2200 MeV,[1] and the situation is further complicated by the fact that more than one resonance occurs in some partial waves. An important problem in strong interaction physics is to understand the dynamical origin of these resonances.

The starting point for most dynamical calculations is the assumption that the scattering matrix is analytic with a cut structure which is known from general theoretical arguments.[2] This allows us to write integral or dispersion relations in the complex plane which define the amplitude in terms of its discontinuities across the cuts. Unfortunately, the partial wave dispersion relations which we obtain in this way are nonlinear integral equations. A general technique introduced by Chew and Mandelstam for reducing the equations to a set of linear integral equations is known as the N/D method. When inelastic effects are important the N/D equations must be written in a matrix form known as the multichannel ND^{-1} equations[3] or one may retain the simple one channel equations by explicitly introducing an inelastic factor η into the equations.[4]

---

*Supported in part by the National Science Foundation.

There are many possible sources of ambiguities which arise in solving the N/D equations. The original partial wave dispersion relations are nonlinear which suggests immediately that the solution of the dispersion relation may not be unique. We can show in a specific example that this is true: the multichannel ND^{-1} equations do not always give the same solution to a problem as one would obtain from the single channel inelastic N/D equations.[5] This nonequivalence of two solutions is related to the presence or absence of S-matrix zeros (or S-matrix poles on an unphysical sheet) in the solution. The partial wave dispersion relations alone are not sufficient to uniquely determine these zeros. Ambiguities of this type were first noted by Castillejo, Dalitz, and Dyson.[6] We will refer to all such ambiguities as CDD effects. The nonequivalence of the two methods of solving the N/D equations implies that the "important" inelastic channels must be explicitly included in a calculation.

Theorems concerning the asymptotic behavior of partial wave amplitudes are clearly important. In particular, a generalized form of Levinson's theorem[7,8] states that the phase shift asymptotically approaches an integer multiple of π where the integer is determined by the number of bound states and S-matrix zeros in a particular partial wave. It would seem that Levinson's theorem is a potentially useful tool to resolve CDD ambiguities. It is dangerous to use it as a guide in practical problems however because, as we shall see, there is no way to tell when energies become large enough to apply asymptotic theorems.[9]

An understanding of CDD effects is crucial to understanding resonance formation. If no CDD effects are present, then a resonance must be due to the "generalized potential" and can be computed by using the single channel equations with inelastic unitarity. On the other hand, if CDD effects are important, then higher mass channels must be explicitly included by using a multichannel formalism. Since the generalized potential terms are not precisely known at present, there is no precise way to tell if CDD effects are responsible for a given resonance. We clearly cannot say that we understand the dynamical origin of a resonance unless we know if CDD effects are responsible for the resonance.

2. PARTIAL WAVE DISPERSION RELATIONS

We will be interested in elastic scattering processes

$$a + b \rightarrow a + b \tag{1}$$

If we let the four momenta of a and b be P_1 and q_1 in the initial state and P_2 and q_2 in the final state, respectively, we can define the invariants

$$s = (P_1 + q_1)^2 = (P_2 + q_2)^2$$
$$t = (P_1 - P_2)^2 = (q_1 - q_2)^2 \qquad (2)$$
$$u = (P_1 - q_2)^2 = (q_1 - P_2)^2$$

where s, t, and u satisfy the constraint

$$s + t + u = 2(m_a^2 + m_b^2) \qquad (3)$$

The elastic scattering is a function of two of these variables, say s and t, and the partial wave amplitude is a function of s alone.

The partial wave scattering amplitude is written as

$$f(s) = \frac{S(s) - 1}{2i\rho(s)} \qquad (4)$$

where $\rho(s)$ is a kinematical factor and S is the scattering matrix. Unitarity (conservation of probability) requires that we write the S-matrix as

$$S = \eta e^{2i\delta} \qquad (5)$$

where

$$0 \leq \eta \leq 1$$

so that (for $s > s_T$)

$$\text{Im } f(s) = \rho|f|^2 + \frac{1 - \eta^2}{4\rho} \qquad (6)$$

The amplitude has cuts and possibly poles in the complex s-plane. The "physical cut" (which we denote by P.C.) is along the real axis for $s \geq s_T = (m_a + m_b)^2$, where s_T is the elastic threshold. There are also cuts in the unphysical region of the s-plane (U. P. C.) which are related by crossing symmetry to the reactions

$$a + \bar{a} \to b + \bar{b} \qquad (t\text{-channel})$$
$$\bar{a} + b \to \bar{a} + b \qquad (u\text{-channel}) \qquad (7)$$

Then by simply using the Cauchy integral formula and deforming the contour we find that

$$f(s) = B(s) + P(s) + \frac{1}{\pi} \int_{\text{P.C.}} \frac{\text{Im } f(s')}{s' - s - i\epsilon} \, ds' \qquad (8)$$

if the asymptotic behavior of the amplitude allows us to write an unsubtracted dispersion relation. Otherwise, we must make subtractions in equation (8). The generalized potential $B(s)$ obtained from the crossed reactions (7) is defined by

$$B(s) = \frac{1}{\pi} \int_{\text{U P.C.}} \frac{\text{Im } f(s')}{s' - s} ds' \tag{9}$$

and $P(s)$ denotes the contribution arising from pole terms due to bound states or elementary particles.

Now if we use unitarity,

$$f(s) = B(s) + P(s) + \frac{1}{\pi} \int_{\text{P.C.}} \frac{(1 - \eta^2) \, ds'}{4\rho(s')(s' - s - i\varepsilon)}$$
$$+ \frac{1}{\pi} \int_{\text{P.C.}} \frac{\rho(s')|f(s')|^2}{s' - s - i\varepsilon} ds' \tag{10}$$

which is a nonlinear integral equation for f if η, B, and P are all known.

3. N/D FORMALISM

1. The N/D formalism is used to convert the nonlinear integral equation of the previous section into a linear integral equation. We assume for the moment that $\eta \equiv 1$ and we write the amplitude in the form

$$f = N/D \tag{11}$$

Now we assume that D has cuts only for physical values of s and N contains only cuts for unphysical s. It follows from unitarity that

$$\text{Im } D(s) = [\text{Im } f^{-1}(s)] \, N(s)$$
$$= -\rho(s)N(s)\theta(s - s_T) \tag{12}$$

If we can write an unsubtracted dispersion relation for D we would have

$$D(s) = -\frac{1}{\pi} \int_{\text{P.C.}} \frac{\rho(s')\theta(s' - s_T)N(s')}{s' - s - i\varepsilon} ds'$$

We are free to normalize D as we choose the amplitude is expressed as a ratio. We can set $D(s_0) = 1$ by making a subtraction at $s = s_0 \leq s_T$.

$$D(s) = 1 - \frac{s - s_0}{\pi} \int_{\text{P.C.}} \frac{\rho(s')\theta(s' - s_T)N(s')}{s' - s_0} \frac{ds'}{s' - s - i\varepsilon} \tag{13}$$

The final expression for f is independent of s_0.

Now we write f as

$$f = B + A^R$$

where A^R has all the physical cuts of f. The quantity $N\text{-}BD \doteq A^R D$ has only the physical cut and satisfies the dispersion relation

$$N(s) - B(s)D(s) = \frac{1}{\pi} \int_{\text{P.C.}} \frac{B(s')\rho(s')N(s')}{s' - s - i\epsilon} ds'$$

If we eliminate D by using equation (13) we obtain an integral equation for N:

$$N(s) = B(s)$$

$$+ \frac{1}{\pi} \int_{\text{P.C.}} \frac{B(s') - [(s - s_0)/(s' - s)]B(s)}{s' - s} \rho(s')\theta(s' - s_T)N(s')\,ds'$$

$$(14)$$

It is easy to show that equation (14) is a Fredholm equation and, hence, has a unique solution if $B(s)$ satisfies the condition that $sB(s) \to$ a constant as $s \to \infty$. (If $\rho(s) \to s$ as $s \to \infty$.)

2. If inelastic effects are present we must modify equations (13) and (14). If the inelasticity is due to the coupling of $n - 1$ two body channels to the elastic channel, then a matrix formalism is appropriate in which the equations for N and D are the same as equations (13) and (14), and B, N, D, and f become $n \times n$ matrices with B and f symmetric. The matrix for ρ is a diagonal matrix of the form $\rho_i(s)\theta(s - s_{T_i})$ where s_{T_i} is the threshold for the ith channel. This method cannot be used to include channels with three or more particles unless we can use an isobar model in which the multiparticle state is approximated by a quasi two-body channel. In any event, the numerical solution of the multichannel ND^{-1} equations is straightforward but time consuming even on fast computers.[10]

3. If the inelastic factor η is known we can derive a set of equations similar to the single channel equation.[4] We note that equation (12) means that $D^*/D = e^{2i\delta}$ for the case $\eta = 1$. For $\eta \neq 1$ we generalize the equations by still requiring that $D^*/D = e^{2i\delta}$ and we find that

$$\text{Im } D(s) = -\frac{2\rho(s)}{1 + \eta(s)} \text{ Re } N(s)\theta(s - s_T)$$

$$\text{Im } N(s) = \frac{1 - \eta(s)}{2\rho(s)} \text{ Re } D(s) \qquad s > s_T \qquad (15)$$

We see that N must now have a cut for physical values of s above the inelastic threshold. The equations which must be solved to obtain N and

D [along with equation (15)] are[4]

$$\frac{2\eta(s)}{1 + \eta(s)} \operatorname{Re} N(s) =$$

$$+ \frac{1}{\pi} \int_{\text{P.C.}} \frac{\bar{B}(s') - [(s - s_0)/(s' - s_0)]\bar{B}(s)}{s' - s} \frac{2\rho(s')}{1 + \eta(s')} \operatorname{Re} N(s') ds' \tag{16}$$

$$D(s) = 1 - \frac{s - s_0}{\pi} \int_{\text{P.C.}} \frac{2\rho(s')}{1 + \eta(s')} \frac{\operatorname{Re} N(s')}{s' - s_0} \frac{ds'}{s' - s - i\epsilon} \tag{17}$$

where

$$\bar{B}(s) = B(s) + \frac{1}{\pi} \mathscr{P} \int_{\text{P.C.}} \frac{1 - \eta(s')}{2\rho(s')} \frac{ds'}{s' - s} \tag{18}$$

and "$\mathscr{P}$" means we must take the principal value integral. The equations allow us to solve for the amplitude, $f = N/D$, for a completely general problem if η is known.

The N/D equations [(13), (14) or (16), (17), and (18)] are a means of ensuring that the amplitude satisfies unitarity for physical s while leaving the unphysical cut terms unchanged.

4. CDD POLES

We have presented a formalism for obtaining a unitary amplitude for a given η and B. The following question now arises: does a knowledge of η and B uniquely determine the scattering amplitude? It is easy to see that the answer is no. All we must do is add pole terms to D[6]

$$D(s) = 1 - \frac{s - s_0}{\pi} \int_{\text{P.C}} \frac{\rho(s')\theta(s - s_T)N(s')}{(s' - s_0)(s' - s - i\epsilon)} ds'$$

$$+ \sum \frac{\lambda_b}{s - s_b}$$

where the s_b lie between s_T and the beginning of the left-hand cut. We can use the same procedure as before to obtain an integral equation for N. The amplitude we obtain still has the same B, the asymptotic behavior is unchanged, and the solution still satisfies unitarity. Adding the poles to D causes the amplitude to have a zero at $s = s_b$. It is interesting to note that if λ_b is small, D will have a zero near s_b. This means that a pole in the amplitude will appear for sufficiently small λ_b regardless of what the potential term B looks like. For large λ_b, D will not necessarily vanish on the physical sheet.

The number of zeros and CDD poles in D determine the asymptotic behavior of the phase shift. A generalized form of Levinson's theorem[7,8] states that

$$\delta(\infty) - \delta(s_T) = \pi(N_{\text{CDD}} - N_P) \qquad (19)$$

where N_P is the number of zeros in D and N_{CDD} is the number of CDD poles in D. In principle Levinson's theorem can be used as a guide to determine if CDD zeros are present. Consider the following examples [let $\delta(s_T) = 0$]:

1. One pole in f.
    (a) If $\delta(\infty) = -\pi$, then $N_{\text{CDD}} = 0$ and the pole is a dynamical bound state.
    (b) If $\delta(\infty) = 0$, then $N_{\text{CDD}} = 1$ and the pole is due to CDD effect.
2. No pole in f, but f has a resonance.
    (a) If $\delta(\infty) = 0$, then $N_{\text{CDD}} = 0$ and the resonance is due to potential term (i.e., the resonance is elastic).
    (b) If $\delta(\infty) = \pi$, then $N_{\text{CDD}} = 1$ and the resonance is due to CDD effect.

If it is necessary to include a CDD pole in D to obtain the correct solution to the partial wave amplitude, then we must have knowledge of λ_b and s_b in addition to $B(s)$. If λ_b and s_b produce a zero in D and hence a bound state, then λ_b and s_b are in turn determined by the mass and coupling constant of the bound state. An appealing idea is to believe that all of the strongly interacting particles result from the dynamics and are not elementary in the sense that they must be explicitly introduced by including CDD effects.[11] We certainly expect this to be true of resonances. We shall see that this is true only if we explicitly include the "important" inelastic channels in a multichannel formalism.

5. S-MATRIX ZEROS

Suppose we have a narrow resonance in a channel at a position below the first inelastic threshold s_{T_2}. There are two possible ways to produce this resonance dynamically: (a) it is due to a complicated potential term B with a long-range repulsion; or (b) it occurs as a quasibound state in a higher mass channel. We can easily produce a sharp bound state by method (b) by using a multichannel formalism. We only have to consider a case in which B_{22} would produce a bound state in channel 2 in the absence of coupling. Now as we "turn on" the coupling we can produce a resonance of arbitrary sharpness in the first channel. For the case $B_{11} \equiv 0$, the Ball–Frazer[12] representation for the phase shift is valid and (for $l = 0$)

$$\delta(s) = -\frac{(s - s_{T_1})^{1/2}}{2\pi} \mathcal{P} \int_{s_{T_2}}^{\infty} \frac{\ln [\eta(s')]}{(s' - s_{T_1})^{1/2}} \frac{ds'}{s' - s} \tag{20}$$

This representation allows us to compute an amplitude (if we use the η found in the multichannel calculation) which is unitary and has same η and B (for $B_{11} = 0$) as we found in the multichannel calculation. It is obvious however that we cannot produce a sharp resonance for $s < s_{T_2}$, since the integral and kinematic factor are varying smoothly in this region. Thus, a knowledge of η alone is not sufficient to uniquely determine the amplitude.

We can demonstrate these ideas explicitly by a two-channel model for $l = 0$ with potential terms

$$B_{ij} = \frac{g_{ij}}{s - s_a}$$

If we choose $s_0 = s_a$ in equation (14) we get $N = B$ and

$$f_{ij} = \sum_k \frac{g_{ik}}{s - s_a} (D^{-1})_{kj}$$

Now if we assume that each channel consists of the scattering of two equal mass spinless particles with nonrelativistic kinematic factors, $\rho_i = (s - s_{T_i})^{1/2}$, we can solve explicitly for D.

$$D_{ij} = \delta_{ij} - g_{ij}\varphi_i$$

$$\varphi_i = -\frac{1}{2}(s_{T_i} - s_a)^{-1/2} + \frac{(s_{T_i} - s_a)^{1/2} - (s_{T_i} - s)^{1/2}}{s - s}$$

For a given matrix g_{ij} we can compute f_{11} and η from the relation

$$\eta^2(s) = 1 - 4\,\rho_1(s)\rho_2(s)|f_{12}(s)|^2\,\theta(s - s_{T_2})$$

Now we can use B_{11} and η as input into the single channel Frye–Warnock equations and compute an amplitude f which we then compare with f_{11}. As we expect, these solutions do not always agree.

To find out what makes the solutions disagree we examine $S_{11} = 1 + 2i\rho_1 f_{11}$. If we solve for the zeros of S_{11}, we obtain a quartic equation in s which we can solve to locate the complex zeros of S_{11}. The same quartic equation also gives the positions of the poles in f_{11} and we must determine which roots correspond to zeros in S_{11} and which roots to poles in f_{11} and on which of the four energy sheets[13] they are on.

For simplicity consider the case $B_{11} = 0$. If B_{12} and B_{22} are "small" we find that f and f_{11} are identical. As B_{22} is made larger there is a critical value of g_{22} where f and f_{11} disagree. At g_{22} greater than this critical value we find that a pair of conjugate zeros in S_{11} have moved onto the physical sheet through the inelastic cut. At the critical value, $\eta = 0$ on the real axis and the single channel equations have a singularity. We call these pairs of zeros CDD zeros with respect to the elastic channel. Now we must write Levinson's theorem as

$$\delta(\infty) - \delta(s_T) = \pi(N_I - N_B)$$

where N_B is the number of bound states and N_I is the number of pairs of zeros which are present in S_{11} and not present in the S-matrix computed from the single channel equations.

6. DISCUSSION

Now the crucial questions become: (a) are these ambiguities due to S-matrix zeros important (i.e., must higher mass inelastic channels be explicitly considered) and (b) can we tell if they are present? We will discuss these questions by considering the well-known P_{33} resonance, N^*_{33} (1238 MeV), in π–N scattering.

Recent phase shift analyses in π–N scattering indicate that the phase shift goes through $90°$ at an incident pion lab kinetic energy E_L of about 190 MeV and stays near $180°$ up to energies of 1300 MeV. Inelastic effects are small in this partial wave up to $E_L = 800$ MeV. The first attempts to compute the N^*_{33} resonance dynamically were done with the notion that the N^*_{33} was produced primarily by forces due to nucleon exchange.[4] The calculation was successful in the sense that a resonance was produced, but the resonant shape did not agree with experiment. More detailed calculations have not improved the agreement with experiment.[15] These were all single channel calculations which attempted to produce the resonance as a result of the potential term B. On the other hand, if S-matrix zeros are present near the N^*_{33} resonant energy, then we would expect to have to include at least one higher mass channel in a multichannel calculation in order to produce the zeros. An attempt to parametrize the N^*_{33} resonance by using a two-channel model which introduced the N^*_{33} as a quasibound state in a higher mass channel was very successful.[16]

Clearly, we cannot say that we understand even the N^*_{33} resonance until we know if it is produced by the potential term or by a higher mass inelastic channel. If $\delta(\infty) = \pi$, then the S-matrix zeros are clearly present, and if $E_L = 1300$ MeV may be considered "asymptotic" then we might conclude that these effects are important. However, as we discussed recently in a simple example,[9] it is impossible to guess where the asymptotic region begins. The potential term is not well known and we might try to approximate it by a series of poles. By adjusting the positions and residues of the poles it is clearly possible to make δ remain close to π over a wide energy range if we include enough poles. The only way to test such a model is to compare the residues with our best estimates from a model calculation and see if they are physically reasonable. There is apparently no way to use Levinson's theorem to resolve this difficulty, since we have no means of determining when a phase shift becomes asymptotic.

So far, we have not been able to *really* calculate details of any of the resonances in $\pi\pi$ on π-N scattering. It is possible however to obtain good fits to the nonresonant low energy π-N phase shifts by solving the partial wave dispersion relations with the appropriate inelasticity.[15] Furthermore, we note that N-N scattering has no resonances and Scotti and Wong[17] obtained really excellent fits in a one-channel calculation. We are also encouraged by the fits obtained to π-N phase shifts by using a simple two-channel model to parametrize the phase shifts and inelasticity.[16] We believe that a realistic calculation of most of the known resonances must be done by explicitly including higher mass channels in a multichannel formalism.

REFERENCES

1. C. Johnson and H. Steiner, Proc. π-N Scattering Conf. Irvine, Dec., 1967 (in press); C. Lovelace, Ibid; P. Bareyre, C. Bricman, and G. Villet, *Phys. Rev.* **165**: 1730 (1968).
2. R. E. Cutkosky, *J. of. Math. Phys.* **1**: 429 (1960).
3. J. D. Bjorken, *Phys. Rev. Letters* **4**: 473 (1960).
4. G. Frye and R. Warnock, *Phys. Rev.* **130**: 478 (1963).
5. M. Bander, P. Coulter, and G. Shaw, *Phys. Rev. Letters* **14**: 207 (1965).
6. L. Castillejo, R. Dalitz, and F. Dyson, *Phys. Rev.* **101**: 453 (1956).
7. J. Hartle and C. Jones, *Ann. Phys.* (N. Y.) **38**: 348 (1966); *Phys. Rev.* **140**: B90 (1965); *Phys. Rev. Letters* **14**: 801 (1965).
8. D. Atkinson, K. Dietz, and D. Morgan, *Ann. Phys.* (N. Y.) **37**: 77 (1966); D. Atkinson and D. Morgan, *Nuovo Cimento* **41**: 559 (1966).
9. P. Coulter, R. Garg, and G. Shaw, *Phys. Rev.* **166**: 1708 (1968).
10. J. Fulco, G. Shaw, and D. Wong, *Phys. Rev.* **138**: B702 (1965).
11. G. F. Chew, *S-Matrix Theory of Strong Interactions*, W. A. Benjamin Inc., New York, 1962.

12. J. Ball and W. Frazer, *Phys. Rev. Letters* 7: 204 (1961).
13. Y. Fujii, *Phys. Rev.* **139B**: 472 (1965).
14. G. Chew, *Phys. Rev. Letters* 9: 233 (1962), F. Low, Ibid 9: 279 (1962).
15. P. Coulter and G. Shaw, *Phys. Rev.* **141**; 1419 (1965); G. Shaw, Proc. π-N Scattering Conf. Irvine, Dec., 1967 (in press).
16. J. Ball. R. Garg, and G. Shaw, *Phys. Rev.* (in press).
17. A. Scotti and D. Wong, *Phys. Rev.* **138**: B145 (1965).

Asymptotic Symmetries and Weak Baryon Form Factors

P. P. DIVAKARAN

TATA INSTITUTE OF FUNDAMENTAL RESEARCH
Bombay, India

1. INTRODUCTION

The idea that many of the approximate internal symmetries of nature may tend to become exact symmetries in certain kinematic limits has been rather fashionable this past year, following the recognition of Costa and Zimerman[1] that it may be used to construct linear combinations of physical amplitudes which converge better than does each amplitude individually. However, the notion that such limits may provide a way of deciding in what sense an apparently badly broken symmetry of nature is a symmetry at all is quite old. In a paper published just before Gell-Mann's work on $SU(3)$, Gell-Mann and Zacharisen[2] studied this question; they conjectured that for propagators and form factors, which are essentially functions of one kinematic variable, say t, this limit is that of infinite t. Similar ideas have also been expressed by Nambu.[3] It is interesting to attempt to trace the history of this idea to the preoccupations of Gell-Mann and Nambu in the period around 1960. It will be recalled that they had, independently, just arrived[4] at a way of looking at the Goldberger–Treiman relation which was far more satisfying than the original derivation of Goldberger and Treiman; the key assumption from which the G–T relation followed almost trivially was seen to be that the matrix element of the divergence of the axial vector isovector current vanished for infinitely large values of t, so that one could write an unsubtracted dispersion relation for this quantity. Now, if the charge corresponding to the axial vector current relevant for the pion decay is a generator of some group and if one assumes that this group is an exact symmetry group *asymptotically* (i.e., for infinite t), then it follows that, asymptotically, the axial vector current has to be conserved. This is precisely the no subtraction assumption. Thus, the G–T relation will be an immediate consequence of an asymptotic symmetry which is generated by, among others, the nonstrange axial charges. The hypothesis of such an asymptotic symmetry will, of course, have other consequences, and it is the purpose of this paper to discuss in a systematic way the implications of some assumed asymptotic symmetries for the weak form factors of baryons.

2. CONSTRUCTION OF SUPERCONVERGENT FORM FACTORS

To illustrate the way in which we shall be using the assumed exactness, asymptotically, of a symmetry, let us consider an internal symmetry group G and two invariant functions $F_1(t)$ and $F_2(t)$ of *one* kinematic variable t, which are such that if G were an exact symmetry group, $F_1(t)$ and $F_2(t)$ will be related through a Clebsch–Gordan coefficient Γ for all t:

$$\frac{F_1(t)}{F_2(t)} = \Gamma \qquad (1)$$

If the symmetry is broken in such a way that asymptotically the exact symmetry relations are preserved, then we have, instead

$$\lim_{t \to \infty} \frac{F_1(t)}{F_2(t)} = \Gamma \qquad (2)$$

It is important to note that for this statement to be meaningful, it is not necessary that $\lim F_i(t)$ should exist individually. What is clear is that the asymptotic behavior of $F_1(t) - \Gamma F_2(t)$ is certainly better than that of $F_1(t)$ or $F_2(t)$ separately. In particular, if a power series expansion exists for F_1 and F_2 [or for $F_1(t)/t^N$ and $F_2(t)/t^N$ for some suitable N] and if $F_1(t)$ [and hence $F_2(t)$] grows as t^α in the asymptotic limit, then $F_1(t) - \Gamma F_2(t)$ grows as t^β where $\beta \leq \alpha - 1$. Let us now assume that $F_1(t)$ and $F_2(t)$ satisfy dispersion relations. If they approach zero as $t \to \infty$, then $F_1(t) - \Gamma F_2(t)$ satisfies a superconvergence relation:

$$\int \text{Im} \, [F_1(t) - \Gamma F_2(t)] \, dt = 0 \qquad (3)$$

Once we reach this stage, we shall be forced to fall back upon the standard (and often not very satisfactory) way of evaluating sum rules such as equation (3), namely, the pole approximation.

This way of arriving at a superconvergence sum rule from considerations of asymptotic symmetry is due to Costa and Zimerman,[1] who started from the assumption that scattering amplitudes satisfy $SU(3)$ relations in the forward direction and in the limit of s, the center-of-mass energy squared, approaching infinity. However, since scattering amplitudes in general depend on two independent kinematic variables, it is not clear what the "asymptotic region" is for them. This ambiguity is absent for 2- and 3-point functions such as propagators and form factors. For propagators, the hypothesis of a suitable asymptotic symmetry *and* a suitable asymptotic behavior[5] leads to spectral function sum rules* of the Weinberg type[6]; and a similar method was used by Pandit[8] to study the effect of symmetry breaking on the K_{l3} form factor F_+.

In our efforts to understand the effects of symmetry breaking on weak baryon form factors, we shall follow essentially the same methods. We first consider the chiral $SU(3) \otimes SU(3)$ as a possible approximate symmetry in the sense that it is exact asymptotically, assuming that $SU(3)$ itself is exact for all t. This leads us to a rough understanding of the effect of symmetry breaking on the axial form factors, in particular the D/F ratio. Next, the possible effects of $SU(3)$ violation are studied assuming $SU(3)$ to be an asymptotic symmetry. Within the approximations we make we shall find that the effect of $SU(3)$ breaking is experimentally unmeasurable—a redefinition of the Cabibbo angles (whose "unrenormalized" value is anyhow not experimentally accessible) is all that results. This would mean that for all practical purposes, $SU(3)$ is an exact symmetry of the weak interactions of baryons—the baryons are exact octet states, and the weak currents transform exactly like appropriate members of two octets. This also provides a justification for our assumption that $SU(3)$ is a good symmetry when looking at the effects of the violation of $SU(3) \otimes SU(3)$.

3. CHIRAL $SU(3) \otimes SU(3)$ AS AN ASYMPTOTIC SYMMETRY

We first write the most general form of the matrix elements of the $\Delta S = 0$ isovector currents between nucleon states. Suppressing ines-

---

*It is impossible to give here a complete account of the work on spectral function sum rules. A reasonably comprehensive survey is attempted in Ref. 7.

sential normalization factors we have, for instance,

$$\langle n, k|V_\mu^-(0)|p, k'\rangle = \bar{u}_n(k)\{\gamma_\mu F_1^f(t)$$
$$+ (k - k')_\nu \sigma_{\mu\nu} [F_2^f(t) + F_2^d(t)]\} u_p(k') \qquad (4)$$

$$\langle n, k|A_\mu^-(0)|p, k'\rangle = \bar{u}_n(k)\gamma_5\{\gamma_\mu [G_1^f(t) + G_1^d(t)]$$
$$+ (k - k')_\mu [G_2^f(t) + G_2^d(t)]\} u_p(k') \qquad (5)$$

where V_μ^- and A_μ^- are the vector and axial vector charge-lowering currents, $t = (k - k')^2$, and the superscripts on the form factors indicate the appropriate symmetry (F and D) type of the coupling. We have also made use of the information that the Dirac form factor F_1 is purely F-type. It is, of course, well known that by far the dominant contributions to the baryon leptonic decay matrix elements comes from the "direct" form factors F_1 and G_1, and a knowledge of these essentially determines, within $SU(3)$, all the baryon leptonic decays.

To relate F to G, it is necessary to know the representations of $SU(3) \otimes SU(3)$ to which the baryons and the currents belong, asymptotically. For the currents, we assume the validity of the commutation relations of Gell-Mann,[9] from which it follows that if the chiral group is a reasonably good one, V_μ^α and A_μ^α transform like $[(1,8)$ and $(8,1)]$. The simplest possibility for the baryons is that they belong to either $[(8, 1) \pm (1, 8)]$ or $[(3, 3^*) \pm (3^*, 3)]$.

That the chiral symmetry is a badly broken one (if indeed there is any sense in thinking of it as a useful symmetry at all) is clear from such facts as the very different masses of the vector and axial vector mesons and the discrepancy in the D/F ratios in the form factors F_1 and G_1. If the baryons belonged to $[(8, 1)$ and $(1, 8)]$, then chiral symmetry will predict that both F_1 and G_1 are purely F-type and that they are equal:

$$\left.\begin{array}{l} G_1^f(t) = F_1^f(t), F_1^d(t) = G_1^d(t) = 0 \\ F_2^f(t) = F_2^d(t) = G_2^f(t) = G_2^d(t) = 0 \end{array}\right\} \text{ for all } t, \qquad (6)$$

whereas if they belong to $[(3, 3^*)$ and $(3^*, 3)]$, one has

$$\left.\begin{array}{l} G_1^d(t) = F_1^f(t), G_1^f(t) = F_1^d(t) = 0 \\ F_2^f(t) = F_2^d(t) = G_2^f(t) = G_2^d(t) = 0 \end{array}\right\} \text{ for all } t \qquad (7)$$

Neither of these possibilities is quite near truth. Apart from the fact that $G_1^f(0) + G_1^d(0) = 1.18 F_1^f(0)$, experiment clearly shows that $G_1^d(0) \approx (1.5 - 2)G_1^f(0)$. Our assumption now is that in spite of this, the matrix elements (or the form factors) are related by the appropriate Clebsch–

Gordan coefficients of $SU(3) \otimes SU(3)$ in the limit of infinite t, with the baryons transforming asymptotically as $[(1, 8)$ and $(8, 1)]$. Thus, we have equations (6) holding not for all t, but only in the limit of infinite t. If we further assume that the form factors are individually convergent, then we get the result

$$\int_0^\infty \text{Im} \, [G_1^f(t) - F_1^f(t)] \, dt = 0 \tag{8}$$

The only poles we know that can contribute to this sum rule are the $A_1(1080)$ in G_1^f and the ρ in F_1^f. Neglecting all other contributions, we write

$$F_1^f(t) = \frac{c_\rho^f m_\rho^2}{m_\rho^2 - t} \tag{9}$$

$$G_1^f(t) = \frac{c_{A_1}^f m_{A_1}^2}{m_{A_1}^2 - t} \tag{10}$$

Equation (8) then leads to

$$G_1^f(0) = F_1^f(0) \left(\frac{m_\rho^2}{m_{A_1}^2} \right) \tag{11}$$

This result is quite interesting for a variety of reasons [assuming, of course, that our truncation of (8) is sensible]: it shows that even though asymptotically G_1 is purely F-type and is equal to F_1, at $t = 0$, there is a very large violation of the chiral symmetry. Secondly, one sees that the smallness of the observed deviation of $G_1(0)/F_1(0)$ $(G_i \equiv G_i^f + G_i^d)$ from 1 is not a true measure of how badly the symmetry is broken for small t. Finally, we have correlated through equation (11) two manifestations (the D/F ratio and the meson masses) of this rather large symmetry breaking.

Numerically, we obtain from equation (11),

$$G_1^f(0) = 0.5 \, F_1^f(0) \tag{12}$$

This number, along with the values of the axial coupling constant $G_1(0) \, [= +1.18 \, F_1^f(0)$, in our notation] and the axial Cabibbo angle θ_A, parametrizes the axial vector matrix elements involved in the weak decays of baryons completely (ignoring the small corrections due to the induced form factors and the t dependence of the form factors). The best experimental determination of $G_1^f(0)/F_1^f(0)$ comes from the β-decays of Λ and Σ^-. $SU(3)$ predicts their ratio to be

$$\frac{\Gamma(\Lambda \to pe^-\bar{\nu})}{\Gamma(\Sigma^- \to ne^-\bar{\nu})} = 0.26 \, \frac{\frac{3}{2}(F_1^f)^2 \sin^2 \theta_V + \frac{1}{2}(G_1^d + 3G_1^f)^2 \sin^2 \theta_A}{(F_1^f)^2 \sin^2 \theta_V + 3(-G_1^d + G_1^f)^2 \sin^2 \theta_A} \tag{13}$$

To the extent that $\theta_A \simeq \theta_V$ this ratio is independent of $\sin \theta$. Substituting $G_1^d = 1.18\, F_1^f - G_1^f$, we can then obtain a value of G_1^f/F_1^f, which may be seen to be extremely insensitive to small deviations from the assumed equality $\theta_A = \theta_V$. Experimentally, one has[10]

$$\frac{\Gamma(\Lambda \to pe^-\bar{\nu})}{\Gamma(\Sigma^- \to ne^-\bar{\nu})} = \frac{8.8 \pm 1.5}{12.5 \pm 1.7}$$

leading to a value*

$$g_A^f \equiv \frac{G_1^f(0)}{F_1^f(0)} = +0.43 \pm 0.05$$

This would also be one of the parameters to be determined from a statistical analysis of all leptonic decays. Thus, for instance, a three parameter[11] fit gives $g_A^f = 0.415 \pm 0.035$, while a four parameter fit[12] leads to $g_A^f \simeq 0.4$. We may thus conclude that our crude model does provide a fairly good approximation to the true state of affairs.

The alternative assumption that baryons belong asymptotically to the representation $[(3, 3^*)$ and $(3^*, 3)]$ will make the combination $G_1^d(t) - F_1^f(t)$ superconvergent [see equation (7)], and will lead to the result

$$G_1^d(0) = 0.5\, F_1^f(0)$$

or

$$g_A^f = 0.7$$

On this basis, we may conclude that if the idea of asymptotic symmetry has any validity, the assignment of the baryons to $[(1, 8)$ and $(8, 1)]$ is favored. Our method is then incapable of providing any information on $G_1^d(0)$, because for this form factor, the asymptotic symmetry merely confirms that it is convergent—it does not provide a nonvanishing "comparison" form factor with the help of which one may form a superconvergent combination.

Irrespective of what representation the baryons belong to, the induced pseuodoscalar form factor $G_2(t) = G_2^f(t) + G_2^d(t)$ can easily be seen to be superconvergent, if indeed the symmetry generators include

---

*Equation (13), being quadratic in G_1^f/F_1^f leads to two values for this quantity, the other being $\simeq 0.90$. This value is generally ignored because it leads to a a value of considerably below that coming from meson decays. Besides, in a least squares fit of all available baryon leptonic decays, the solution we have retained is favored.

the axial charges; for in that case, all axial currents would be asymptotically conserved. In particular,

$$\lim_{t\to\infty} \langle n, k|\partial_\mu A_\mu^-(0)|p, k'\rangle \equiv \lim_{t\to\infty} u_n(k)\gamma_5 u_p(k')[2m_N G_1(t) + t G_2(t)] = 0$$

(14)

Since $G_1(t)$ is assumed to be convergent, it follows that $G_2(t)$ is superconvergent. As long as we are limited to the (π, A_1) pole saturation of the resulting sum rule, it is easy to see that this leads to nothing more than the Goldberger–Treiman relation and a small (because of the relatively large value of $m_{A_1}^2$) correction to the usual pion pole value for $G_2(0)$.[13]

4. $SU(3)$ AS AN ASYMPTOTIC SYMMETRY

In this part we relax the assumption that $SU(3)$ itself is an exact symmetry and consider the consequences of its violation in the same spirit as above—i.e., we assume that the baryons, the vector and axial vector currents all transform as octets as far as matrix elements of these currents at infinite t are concerned. We shall saturate the resulting superconvergent combinations of weak form factors by known poles, and attempt in this way to determine the corrections to the usual Cabibbo theory.

The weak leptonic decays on which we have any sort of information available are listed in Table 1, the first two being $\Delta S = 0$ transitions and the others being $|\Delta S| = 1$ transitions. The expressions (in the limit of exact symmetry) for the corresponding matrix elements of the appropriate components of a general octet current in terms of the usual D and F reduced matrix elements are also given.

Table 1. Expressions for Matrix Elements in $SU(3)$

Decay	$n\to p$	$\Sigma^-\to\Lambda$	$\Lambda\to p$	$\Sigma^-\to n$	$\Xi^-\to\Lambda$
Matrix element	$D + F$	$-\sqrt{\frac{2}{3}}D$	$\frac{1}{\sqrt{6}}D + \sqrt{\frac{3}{2}}F$	$-D + F$	$\frac{1}{\sqrt{6}}D - \sqrt{\frac{3}{2}}F$

In general, both the vector and the axial currents contribute to these decays, and we shall again assume that the value of the matrix element relevant for a decay is given sufficiently well by the corresponding *direct* form factor at $t = 0$—i.e., t-dependence and induced form factors are again ignored.

Since the matrix elements of the vector currents are pure F, exact $SU(3)$ will result in four relations among the five amplitudes, e.g.,

$$\langle p|V_\mu|\Lambda\rangle = \sqrt{\frac{3}{2}} \langle p|V_\mu|n\rangle \qquad (15)$$

or, in our approximation,

$$F_1(\Lambda \to p) = \sqrt{\frac{3}{2}} F_1(n \to p) \qquad (16)$$

in an obvious notation. As before, asymptotic symmetry tells us, instead, that $F_1(\Lambda \to p) - \sqrt{\frac{3}{2}} F_1(n \to p)$ is superconvergent:

$$\int \text{Im}\left[F_1(\Lambda \to p) - \sqrt{\frac{3}{2}} F_1(n \to p) \right] dt = 0 \qquad (17)$$

The only single particle states that contribute to this sum rule are K^* [to $F_1(\Lambda \to p)$] and ρ [to $F_1(n \to p)$]. So we have, from equation (17),

$$G_{\Lambda p K^*} m_{K^*}^2 - \sqrt{\frac{3}{2}} G_{n p \rho} m_\rho^2 = 0 \qquad (18)$$

where the coupling constants are defined, for example, by

$$F_1(\Lambda \to p) = \frac{G_{\Lambda p K^*} m_{K^*}^2}{m_{K^*}^2 - t} \qquad (19)$$

Thus,

$$\frac{F_1(\Lambda \to p, t = 0)}{F_1(n \to p, t = 0)} = \sqrt{\frac{3}{2}} \frac{m_\rho^2}{m_{K^*}^2} \qquad (20)$$

or, in our approximation,

$$\frac{\langle p|V_\mu|\Lambda\rangle}{\langle p|V_\mu|n\rangle} = \frac{m_\rho^2}{m_{K^*}^2} \left[\frac{\langle p|V_\mu|\Lambda\rangle}{\langle p|V_\mu|n\rangle} \right]_{SU(3)} \qquad (21)$$

We can deal with all the decays in this fashion and thus get the interesting result that the effect of the symmetry breaking, in our model, is to multiply the $SU(3)$ value of the matrix element by the inverse squared mass of the vector meson with the appropriate hypercharge:

$$\langle B'|V_\mu(\Delta S = 0)|B\rangle = \frac{1}{m_\rho^2} \langle B'|V_\mu(\Delta S = 0)|B\rangle_{SU(3)}$$

$$\langle B'|V_\mu(|\Delta S| = 1)|B\rangle = \frac{1}{m_{K^*}^2} \langle B'|V_\mu(|\Delta S| = 1)|B\rangle_{SU(3)} \qquad (22)$$

Within the framework of the Cabibbo theory, however, the observable matrix elements are not (22), but rather those of the Cabibbo current which is schematically

$$V_\mu^{\text{cab.}} = \cos\theta V_\mu(\Delta S = 0) + \sin\theta V_\mu(|\Delta S| = 1) \qquad (23)$$

In the limit of exact symmetry, $V_\mu(\Delta S = 0)$ and $V_\mu(|\Delta S| = 1)$ are appropriate members of an octet. Instead of a quantity like the left-hand side of equation (21), what is really observable is only

$$\tan\theta \frac{\langle p|V_\mu|\Lambda\rangle}{\langle p|V_\mu|n\rangle} = \sqrt{\frac{3}{2}}\frac{m_\rho^2}{m_{K^*}^2}\tan\theta \tag{24}$$

Therefore, as far as the matrix elements of the vector current are concerned, our model predicts that the effect of symmetry breaking is simply to "renormalize" the value of the Cabibbo angle to

$$\tan\theta_V = \frac{m_\rho^2}{m_{K^*}^2}\tan\theta \tag{25}$$

Since the unrenormalized Cabibbo angle is not accessible to experiment, the symmetry breaking is unobservable.

It is clear that the axial vector current may be dealt with essentially along the same lines. Since D- and F- type form factors are both involved now, there are only three relations among the five form factors, when symmetry is exact. We may conveniently express all $|\Delta S| = 1$ matrix elements in terms of the two $\Delta S = 0$ ones $n \to p$ and $\Sigma^- \to \Lambda$. For example, we have

$$\langle p|A_\mu|\Lambda\rangle = \sqrt{\frac{3}{2}}\langle p|A_\mu|n\rangle + \langle\Lambda|A_\mu|\Sigma^-\rangle \tag{26}$$

In our model of symmetry breaking, the corresponding superconvergence condition is

$$G_{\Lambda p K_A}m_{K_A}^2 - \left(\sqrt{\frac{3}{2}}G_{np A_1} + G_{\Sigma^-\Lambda A_1}\right)m_{A_1}^2 = 0 \tag{27}$$

where we have assumed the $|\Delta S| = 1$ axial form factor to be dominated by the $K_A(1320)$ meson and the $\Delta S = 0$ one by $A_1(1080)$ and where the symbols G for coupling constant are defined as in equation (19). As before, we can determine the ratio

$$\frac{\langle p|A_\mu^{\text{cab.}}|\Lambda\rangle}{\sqrt{\frac{3}{2}}\langle p|A_\mu^{\text{cab.}}|n\rangle + \langle\Lambda|A_\mu^{\text{cab.}}|\Sigma^-\rangle} = \frac{m_{A_1}^2}{m_{K_A}^2}\tan\theta \tag{28}$$

assuming, following Cabibbo, that the vector and axial Cabibbo angles are the same.

So we see that our conclusions for the axial vector matrix elements are the same as for the vector matrix elements—the effect of symmetry breaking is again simply to renormalize the Cabibbo angle to a value given by

$$\tan\theta_A = \frac{m_{A_1}^2}{m_{K_A}^2}\tan\theta \tag{29}$$

Numerically, $m_{A_1}^2/m_{K_A}^2 = m_\rho^2/m_{K^*}^2$, a result which also follows from considerations of asymptotic symmetry for two-point functions,[5] leading to our final conclusion, that in our model the breaking of $SU(3)$ is completely unobservable in baryon leptonic decays, and that an analysis of the experimental data on the basis of exact $SU(3)$ and with one Cabibbo angle ($\theta_A = \theta_V$) should be satisfactory. It should be mentioned here that the flexibility of the Cabibbo form of universality, with θ unspecified, is essential for this conclusion to be drawn. We may also note that the same result has also been shown to be true in this model for the semileptonic K_{e3} and π_{e3} decays.[8]

Experimentally, this picture of baryon leptonic decays has good support, bearing in mind that available information on some of these decays is extremely meagre. Most analyses of these data are done assuming that $SU(3)$ symmetry is exact. Thus the fit of Courant et al.[14] and the later one of Carlson[11] showed that the Cabibbo theory in its original form with $\theta_A = \theta_V$ describes these decay processes satisfactorily. The analysis of Brene et al.[12] allows for what they call a certain amount of symmetry breaking by allowing θ_A and θ_V to be independent parameters. This procedure is consistent with our model, where we have seen that the effect of symmetry breaking is to change the effective value of θ_A and θ_V from their unrenormalized values, even though Brene et al. were led to this procedure by the Ademollo–Gatto theorem which says that to the first order, $SU(3)$ violation leaves the matrix elements of the vector currents unchanged. In our model, this is far from being the case—but then ours is in no sense a first-order calculation.

Brene et al. find the best fit to correspond to $\theta_V = 0.21$ and $\theta_A = 0.27$, which is to be compared with our prediction $\theta_A = \theta_V$. But the fact that one can get good fits to the data with $\theta_A = \theta_V$ already indicates that the ratio θ_A/θ_V is sensitive to the input. This can also be seen from equation (13), which can be used to determine $\tan^2 \theta_A/\tan^2 \theta_V$ as a function of g_A^f. For the experimentally measured values of $\Gamma(\Lambda \to p)$ and $\Gamma(\Sigma^- \to n)$, $\tan^2 \theta_A/\tan^2 \theta_V$ is singular at $g_A^f \simeq 0.36$ and varies strongly in a range of g_A^f which is acceptable experimentally. Our final conclusion is then that allowing for the inevitable crudeness of our model, the idea that symmetries such as $SU(3) \otimes SU(3)$ and $SU(3)$ are broken in such a way that they are asymptotically exact for form factors does provide a correct description of baryon leptonic decays in our present state of knowledge about them.

REFERENCES

1. G. Costa and A. H. Zimerman, *Nuovo Cimento* **46A**: 198 (1967).
2. M. Gell-Mann and F. Zachariasen, *Phys. Rev.* **123**: 1065 (1961). [See also M. Gell-Mann, "Weak Interactions of Strongly Interacting Particles" (Lectures at the 1961 Bangalore Summer School), Tata Institute of Fundamental Research, Bombay, 1961].
3. Y. Nambu, "Proceedings of the Istanbul Summer School of Theoretical Physics (1962)," Ed. F. Gürsey, Gordon and Breach, New York and London.
4. Y. Nambu, *Phys. Rev. Letters* **4**: 380 (1960); J. Bernstein, M. Gell-Mann and L. Michel, *Nuovo Cimento* **16**: 560 (1960).
5. T. Das, V. S. Mathur, and S. Okubo, *Phys. Rev. Letters* **18**: 761 (1967).
6. S. Weinberg, *Phys. Rev. Letters* **18**: 507 (1967).
7. P. P. Divakaran, "Proceedings of the 10th Symposium on Cosmic Rays, Elementary Particle Physics and Astrophysics," Aligarh (December, 1967), to be published by the Department of Atomic Energy, Bombay.
8. L. K. Pandit, *Phys. Rev. Letters* **19**: 263 (1967).
9. M. Gell-Mann, *Phys. Rev.* **125**: 1067 (1961); *Physics* **1**: 63 (1964).
10. A. H. Rosenfeld *et al.*, UCRL Peport 8030 (September 1967).
11. C. E. Carlson, *Phys. Rev.* **152**: 1433 (1966).
12. N. Brene, L. Veje, M. Roos, and C. Cronström, *Phys. Rev.* **149**: 1288 (1966).
13. T. D. Lee and C. S. Wu, *Ann. Rev. Nuclear Sci.* **15**: 381 (1965).
14. H. Courant, H. Filthuth, P. Franzini, A. Minguzzi-Ranzi, A. Segar, R. Engelmann, V. Hepp, E. Kluge, R. A. Burnstein, T. B. Day, R. B. Glasser, A. J. Herz, B. Kehoe, B. Sechi-Zorn, N. Seeman, G. A. Snow, and W. Willis, *Phys. Rev.* **136**: B1791 (1964).

Implications of Current Algebra for η Decay— A Summary*

R. RAMACHANDRAN† AND ADITYA KUMAR†

TATA INSTITUTE OF FUNDAMENTAL RESEARCH
Bombay, India

INTRODUCTION

In calculations involving pions, the use of equal time commutation relations for the currents together with PCAC (Partial Conservation of Axial Vector Current) principle for the pion field operator has had some success over the last couple of years. Most spectacular among these is the calculation of Adler[1] and Weissberger[2] giving the weak axial vector renormalization g_A in terms of the total crosssection in $\pi-N$ scattering. Subsequently, the current algebra was found successful in relating various leptonic K-decay processes[3] and in giving some details of nonleptonic decays.[4] Hara and Nambu[4] applied these techniques to successfully predict the energy spectrum of the unlike pion in the $K \to 3\pi$ decay. There is a lot of similarity between $\eta \to 3\pi$ and $K \to 3\pi$ decays, and it is natural to expect that similar mechanism explain both processes. However, the current algebra techniques that were so successful in K decays have not had a similar effect in η decays. We shall see that some of the difficulties will be traced to the ambiguity in the various extrapolations possible from the soft pion limit (where the current algebra makes definite predictions) to the physical pions. In this talk we shall review

---

*Presented at the Sixth Anniversary Symposium, The Institute of Mathematical Sciences, Adyar, Madras, India, January, 1968.

†Present address: Indian Institute of Technology, Kanpur, India.

briefly the various approaches and then suggest an extrapolation procedure that we think best explains the η-decay process.

The decay process is G violating and so far there is no evidence that the charge conjugation symmetry is violated. This would normally suggest that the decay is electromagnetic in origin. If we take the usual current–current form, then up to the second order, the final three-pion state should be in pure $I = 1$ state. But on the basis of such a decay Hamiltonian, it is not possible to understand the various branching ratios. For instance, $\eta \to 2\gamma$ and $\eta \to 3\pi$ are both second order in e, and on the basis of phase space $\eta \to 2\gamma$ should be much more dominant than the $\eta \to 3\pi$ mode. But experimentally the rates are comparable. We need to find a mechanism that explains the enhancement of the $\eta \to 3\pi$ mode. Alternatively, one may give up the usual form altogether and consider that the decay is mediated through a phenomenological scalar isovector density, very much like the weak interactions.[5] For example, in the quark model this would have the form $H = \bar{q}(x)\lambda_3 q(x)$. We shall use both forms of Hamiltonian together with the Gell-Mann equal time commutation relations in the following discussion.

SCALAR DENSITY MODEL

Let us write the matrix element for the decay $\eta \to \pi_\alpha(q_1)$, $\pi_\beta(q_2)$, $\pi_\gamma(q_3)$, (ignoring the normalization etc.):

$$M^{\alpha\beta\gamma}(q_1, q_2, q_3) = \frac{1}{c^3}(q_1^2 - \mu^2)(q_2^2 - \mu^2)(q_3^2 - \mu^2)\int dx\,dy\,dz\,e^{i(q_1 x + q_2 y + q_3 z)}$$
$$\times \langle 0|T\{D^\alpha(x), D^\beta(y), D^\gamma(z), H(0)\}|\eta\rangle \qquad (1)$$

where $D^\alpha(x) = \partial_\mu A_\mu^\alpha(x)$ and the PCAC is assumed through

$$\phi^\alpha(x) = \frac{1}{c_\alpha}\partial_\mu A_\mu^\alpha(x) \qquad c_\pi = 0.69 \pm 0.04\,m_\pi^3 \qquad (2)$$

Let us further define a charge:

$$Q^\alpha(x_0) = \int d^3x\,A_0^\alpha(x) \qquad (3)$$

By scalar-density model, we understand that the Hamiltonian is given by

$$H(x) = u^3(x) = \bar{q}(x)\lambda_3 q(x) \qquad (4)$$

At equal times we have the commutation relations

$$[Q^\alpha(x_0), u^3(x)] = -d_{3\alpha\beta}v^\beta(x); \quad v^\beta(x) = \bar{q}(x)\gamma_5\lambda_\beta q(x) \qquad (5)$$

and

$$[Q^\alpha(x_0), v^\beta(x] = -d_{\alpha\beta\gamma}u^\gamma(x) \tag{6}$$

We will also encounter commutation relation of axial charges with the divergences and these would signify s-wave π-π correlations and correspond to the elusive σ meson if it does exist. Experimentally, the π-π interaction is all but negligible in the $I = 2$ channel. Therefore, we will retain only the isoscalar part, and have as σ terms

$$[Q^\alpha(x_0), D^\beta(x)] = \delta_{\alpha\beta}\sigma(x) \tag{7}$$

Recalling that the current algebra techniques yield rather small values of scattering lengths for s-wave π-π interaction,[6] we may be justified in ignoring the σ terms altogether, as we shall do presently in this model.

We may now contract one of the three pions and let its four-momentum to zero:

$$M^{\alpha\beta\gamma}(0, q_2, q_3) = -\frac{\mu^2(q_2^2 - \mu^2)(q_3^2 - \mu^2)}{c^3} \int dx\, dy\, dz\, e^{i(q_2 y + q_3 z)}$$
$$\times \langle 0|T\{\partial_\mu A_\mu^\alpha(x), D^\beta(y), D^\gamma(z), u^3(0)|\eta\rangle \tag{8}$$

Using the identity

$$\partial_\mu T\{A_\mu^\alpha(x), D^\beta(y), D^\gamma(z)u^3(0)\} = T\{\partial_\mu A_\mu^\alpha, D^\beta, D^\gamma, u^3\}$$
$$+ \delta(x_0 - y_0)T\{[A_0^\alpha(x), D^\beta(y)], D^\gamma, u^3\}$$
$$+ \delta(x_0 - z_0)T\{[A_0^\alpha(x), D^\gamma(z)], D^\beta, u^3\}$$
$$+ \delta(x_0)T\{[A_0^\alpha(x), u^3(0)]D^\beta, D^\gamma\} \tag{9}$$

and the appropriate commutation relations, we get

$$M^{\alpha\beta\gamma}(0, q_2, q_3) = \frac{\mu^2(q_2^2 - \mu^2)(q_3^2 - \mu^2)}{c^3} \int dy\, dz\, e^{i(q_2 y + q_3 z)}$$
$$\times \{d_{3\alpha\lambda}\langle 0|T\{D^\beta(y)D^\gamma(z)v^\lambda(0)\}|\eta\rangle$$
$$+ \delta_{\alpha\beta}\langle 0|T\{\sigma(y)D^\gamma(z)u^3(0)\}|\eta\rangle$$
$$+ \delta_{\alpha\gamma}\langle 0|T\sigma(z)D^\beta(y)u^3(0)\}|\eta\rangle\} \tag{10}$$

Now if we ignore the σ terms as unimportant, then the matrix element is proportional to $d_{3\alpha\lambda}$ and thus is nonvanishing only when $\alpha = 3$. This implies that whenever either of the charged pions is made soft (i.e., $q \to 0$) the matrix element vanishes. This, together with the assumption that the matrix elements depends only linearly on the energy of the pions (which is suggested by the experimental spectrum), is enough to give the slope of the spectrum. In terms of the usual Dalitz variables

$$x = \frac{\sqrt{3}\,(E_+ - E_-)}{(m - 3\mu)} \quad \text{and} \quad y = \frac{3E_0 - m}{(m - 3\mu)}$$

where m and μ are masses of η and π, respectively, in the linear form, the decay matrix element reads

$$M(x, y) = 1 + by \tag{11}$$

when normalized to unity at the center of the Dalitz plot. When the charged pion momentum vanishes, $(E_+ \text{ or } E_-) = 0$ and $E_0 = (E_- \text{ or } E_+) = m/2$ This would correspond to

$$x = \frac{\pm\sqrt{3}\,m}{m - 3\mu} \qquad y = \frac{m}{2(m - 3\mu)}$$

The condition that

$$M\left(x = \frac{\pm\sqrt{3}\,m}{m - 3\mu}, y = \frac{m}{2(m - 3\mu)}\right) = 0 \tag{12}$$

would then imply

$$b = -\frac{2(m - 3\mu)}{m} = -0.49 \tag{13}$$

This, indeed, compares very well with the experimental number[7]

$$b = 0.478 \pm 0.038 \tag{14}$$

ELECTOROMAGNETIC HAMILTONIAN

Secondly, we shall now consider η decay as arising from the usual current–current e.m. Hamiltonian. Recalling that only the isovector part will contribute to the decay, we will have

$$H^3(0) = \int dx D^{\mu\nu}(x) j_\mu^{(3)}(x) j_\nu^{(8)}(0) \tag{15}$$

The appropriate commutation relations at equal times are

$$[Q^\alpha(0), H^3(0)] = if_{3\beta\alpha} \int dx D^{\mu\nu}(x) A_\mu^\beta(x) j_\nu^{(8)}(0)$$

$$= if_{3\beta\alpha} H'^\beta(0) \tag{16}$$

and

$$[Q^\alpha(0), H'^\beta(0)] = if_{\alpha\beta\gamma} H^\gamma(0) \tag{17}$$

On contracting one of the pions and using the identity (9), we will have instead of equation (10) (ignoring the σ terms)

$$M^{\alpha\beta\gamma}(0, q_2, q_3) = \frac{1}{c^3} \mu^2 (q_2^2 - \mu^2)(q_3^2 - \mu^2) i f_{3\lambda\alpha} \int dy \, dz \, e^{i(q_2 y + q_3 z)}$$

$$\times \langle 0|T\{D^\beta(y)D^\gamma(z)H'^\lambda(0)\}|\eta\rangle \qquad (18)$$

The isospin structure would indicate that the equation (18) be completely antisymmetric in the isospin indices β, γ, λ, whereas the momentum conservation would force symmetry between y and z. Thus, if at this unphysical point, pions obey Bose statistics, then equation (18) must vanish. Note that this happens whichever pion is made soft and not just the charged ones. If, further, the matrix element has a linear form, we are forced to have an identically vanishing matrix element. This was first observed by Sutherland.[8]

A way out of this dilemma is given by Bardeen *et al.*[9] Briefly, it consists in the observation that the matrix element in the presence of σ terms, depends, in addition to the energy variables, on the off mass-shell pion mass. Notice that when the σ terms are not neglected [the last two terms of equation (10)]. Equation (18) vanishes only when $q_2^2 \rightarrow \mu^2$ and $q_3^2 \rightarrow \mu^2$, in addition to $q_1 = 0$. The most general form for the matrix element, consistent with Bose statistics is then given by

$$M(x, y) = A(1 + ay + cz_0) \qquad z_0 = q^2 \qquad (19)$$

for neutral pion, still retaining only linear terms. When π^0 is made soft, keeping π^+ and π^- on the mass shell, we get

$$1 - \frac{m}{m - 3\mu} a = 0 \qquad (20a)$$

Similarly, when π^+ or π^- is made soft, we have

$$1 + \frac{m}{2(m - 3\mu)} a + c\mu^2 = 0 \qquad (20b)$$

Solving (20a) and (20b), we get

$$M(x, y) = A\left(1 + \frac{m - 3\mu}{m} y - \frac{3}{2\mu^2} z_0\right) \qquad (21)$$

However, on the mass shell, this has precisely the same form as the one obtained in the scalar Hamiltonian model.

This explanation, however, crucially depends on the importance of σ terms. This is surprising in view of the small $\pi-\pi$ scattering lengths obtained from similar current algebra arguments. This also exposes the arbitrariness in the extrapolation procedure from soft pion limit inherent in the current algebra techniques. We shall now make use of this arbitrariness and attempt to give yet another extrapolation in which again σ terms will be considered absent.

QUADRATIC TERMS IN THE MATRIX ELEMENT

The method stems from the observation that the Dalitz plot for η decay does not rule out the presence of a small term quadratic in the energy variable. The most general matrix element for η decay, up to the second order in energy is given by

$$M(x, y) = 1 + by + a_s(x^2 + y^2) + a_m(x^2 - y^2) \qquad (22)$$

The term proportional to $(x^2 + y^2)$ and the constant are associated with the completely symmetric three-pion state, while the mixed symmetric state gives rise to terms proportional to y and $x^2 - y^2$. Since the matrix elements vanishes at the soft pion limits, (i.e., at $x = \pm\sqrt{3}\,m/(m - 3\mu)$, $y = m/2(m - 3\mu)$ and $x = 0, y = -m/(m - 3\mu)$) we will find, by substituting in equation (22)

$$a_s = -\left(\frac{m - 3\mu}{m}\right)^2 = -0.061 \qquad (23a)$$

and

$$b + \frac{m}{m - 3\mu}\,a_m = 0 \qquad (23b)$$

Without any additional assumption, the current algebra cannot give the values of the coefficients b and a_m. One may, however, use the experimental slope at the centre of the Dalitz plot (which incidentally is almost the same as that obtained on the basis of the scalar density Hamiltonian) and use equation (23b) to estimate the magnitude of the quadratic terms needed on the basis of the current algebra techniques. With $b = -0.50$, a_m is 0.123. Then, we have

$$M(x, y) = 1 - 0.50y - 0.061(x^2 + y^2) + 0.123(x^2 - y^2) \qquad (24)$$

Equation (24) clearly shows that quite small coefficients are sufficient to avoid the catastrophe of a vanishing matrix element vis-a-vis current algebra. A comparison with the the the experimental π^0 energy spectrum and and that implied by equation (24) indicates it to be as good as the linear matrix element. Further if the Dalitz plot were divided into three radial zones of equal area, the distribution of events in the inner; middle; and the outer zones are in the ratio 1.00: 1.07 $\pm$ 0.07: 0.93 $\pm$ 0.07. Whereas a linear matrix element would imply an increase as we go out from the center (typically, for the parameter in equation 14, it is 1.00: 1.04: 1.07), the above matrix element gives a distributions in the ratio 1.00: 1.00: 1.00. Indeed, the quadratic matrix element derived from the current algebra agrees well with the Dalitz plot.

EFFECTIVE LAGRANGIAN

The uncertainty one encounters in applying current algebra to η processes may have its origin elsewhere. For example, in the effective Lagrangian approach the PCAC principle (that plays a central role in all current algebra calculation) is capable of being given an operational definition only for processes not involving η. We will give here briefly the arguments that lead to this conclusion. Cronin[10] has given a method for constructing an effective Lagrangian for a nonet of pseudoscalar mesons, that incorporates the chiral $U(3) \times U(3)$ symmetry (Gell-Mann current commutation relations). The effective Lagrangian implies that only the lowest order results are considered and the higher order terms do not have any significance. The primary ingredient is the construction of meson-coupling matrix M as a function of the 3×3 Hermitian pseudoscalar-meson matrix Φ. This matrix M is defined to transform according to the representation $(3, 3^*)$ of $U(3) \times U(3)$. The only constraints, this function has are

$$M^\dagger(\Phi)M(\Phi) = I \quad \text{and} \quad M^\dagger(\Phi) = M(-\Phi) \tag{25}$$

The pseudoscalar mesons do not belong to a linear representations of this group. A power series expansion of M in terms of Φ, subject to equation (25) has the form

$$\begin{aligned} M(\Phi) = 1 + 2if\Phi + 2(if\Phi)^2 + a_3(if\Phi)^3 \\ + 2(a_3 - 1)(if\Phi)^4 + \cdots \end{aligned} \tag{26}$$

where the parameter f is later identified with the pion decay constant. The phenomenological Lagrangian is constructed by writing a chiral invariant kinetic term

$$\mathscr{L}_k = \frac{1}{8f^2} \text{Tr}(\partial_\mu M^\dagger \partial_\mu M) \tag{27}$$

and adding a symmetry breaking mass term. The mass term is so constructed that the second-order terms correspond to the physical meson masses. Accordingly,

$$\mathscr{L}_m = \frac{1}{8f^2} \text{Tr}\{(a + b\lambda_8)(M + M^+)\}$$

$$+ \frac{c}{64f^2} \text{Tr}\{\lambda_0(M - M^+)\}\text{Tr}\{\lambda_0(M - M^+)\}$$

$$+ \frac{d}{64f^2} \text{Tr}\{\lambda_0(M - M^+)\}\text{Tr}\{\lambda_8(M - M^+)\}$$

$$+ \frac{e}{64f^2} \text{Tr}\{\lambda_8(M - M^+)\}\text{Tr}\{\lambda_8(M - M^+)\} \qquad (28)$$

where

$$a = \frac{1}{3}(2m_\kappa^2 + m_\pi^2)$$

$$b = \frac{2}{\sqrt{3}}(m_\pi^2 - m_\kappa^2)$$

$$c = m_{\lambda_0}^2 \cos^2 \lambda + m_\eta^2 \sin^2 \lambda - a$$

$$d = 2 \sin \lambda \cos \lambda(m_{\lambda_0}^2 - m_\eta^2) - 2\sqrt{\frac{2}{3}} b$$

$$e = m_{\lambda_0}^2 \sin^2 \lambda + m_\eta^2 \cos^2 \lambda - \frac{1}{3}(4m_\kappa^2 - m_\pi^2)$$

λ stands for octet singlet mixing. We shall assume that Gell–Mann–Okubo mass formula is valid and set $e = 0$. Given this Lagrangian it is possible to construct the axial vector current. Since the mass term breaks chiral symmetry, the divergence of axial vector current will have nonvanishing contributions from such terms. We find that

$$-\partial_\mu A_\mu^p = \frac{i}{16f^2} \text{Tr}[(a + b\lambda_8)\{(M - M^+), \lambda_p\}_+]$$

$$+ \frac{ic}{64f^2} \text{Tr}[\lambda_0(M - M^+)] \times \text{Tr}[\lambda_0\{(M + M^+), \lambda_p\}_+]$$

$$+ \frac{id}{128f^2} \text{Tr}\{[\lambda_0(M - M^+)] \times \text{Tr}[\lambda_8\{(M + M^+), \lambda_p\}_+]$$

$$+ \text{Tr}[\lambda_0\{(M + M^+), \lambda_p\}] \times \text{Tr}[\lambda_8(M - M^+)]\} \qquad (29)$$

We may now expand equation (29) and find up to third order in ϕ

$$\partial_u A_\mu^p = \frac{m_p^2}{f\sqrt{2}}\,\phi_p - \frac{a_3 f}{16\sqrt{2}}\,\phi_a\phi_b\phi_c\{\mathrm{Tr}[(a + b\lambda_8)\{\lambda_a\lambda_b\lambda_c, \lambda_p\}_+]$$
$$+ 2c\delta_{p0}\mathrm{Tr}[\lambda_0\lambda_a\lambda_b\lambda_c] + d\delta_{p0}\mathrm{Tr}[\lambda_8\lambda_a\lambda_b\lambda_c]$$
$$+ d\delta_{p8}\mathrm{Tr}[\lambda_0\lambda_a\lambda_b\lambda_c]\}$$
$$- \frac{f}{8\sqrt{2}}\,\phi_8\phi_a\phi_b\{2c\mathrm{Tr}[\lambda_0\{\lambda_a\lambda_b, \lambda_p\}_+] + d\,\mathrm{Tr}[\lambda_8\{\lambda_a\lambda_b, \lambda_p\}_+]\}$$
$$- \frac{fd}{8\sqrt{2}}\,\phi_8\phi_a\phi_b\mathrm{Tr}[\lambda_0\{\lambda_a\lambda_b, \lambda_p\}_+] + \phi^5 \text{ terms} + \cdots \qquad (30)$$

We notice that if a_3 is set zero (which is a free adjustable parameter so far) then we recover PCAC in the conventional form except for the last two terms in equation 30. These terms will contribute only if η or x_0 fields are involved. Thus, in all processes involving up to four mesons PCAC is exact once we ignore the η and x_0 fields. Cronin has worked out the consequences of this Lagrangian and he obtains for π-π, π-K and K-K scattering and K meson decays results in agreement with the usual current algebra methods and are roughly in accord with the experiment. On extending his model to η processes and in particular to $\eta \rightarrow 3\pi$ decay, we get results not in agreement with either the current algebra procedures, we have discussed or the experiment. We believe that this is due to the arbitrariness we now have in the absence of a strict PCAC, unlike the kaon processes for which the Lagrangian had a unique form.

FINAL-STATE INTERACTIONS

Like all other current algebra results, the matrix element for η decay discussed so far has been purely real. One is not sure whether this means that the final-state interactions have been deliberately thrown out or that our procedure is designed to give only the magnitude of the matrix element and is to be supplemented with a phase, calculated by considering the final-state interactions in the usual sense. Phenomenologically it is possible to obtain a good fit for the Dalitz plot by including fairly large imaginary parts in the coefficients of the energy variable. Such complex matrix elements have an additional property in that they are capable reducing the branching ratio $R = \Gamma(\eta \rightarrow \pi^0\pi^0\pi^0)/\Gamma(\eta \rightarrow \pi^+\pi^-\pi^0)$ from the value of 1.63 (for real coefficients) to 0.9 (if the matrix element is quadratic; if it is cubic the lower limit for R is 0.5) still maintaining $\Delta I = 1$ rule.[11] The experimental value for R lies in the region of 1.3 to

1.55. Should it turn out to be close to 1.3, it will be unavoidable that we require some imaginary part as well as quadratic terms in the matrix element for a fit. This in turn should suggest a strong final state interaction. The treatment of f.s.i. in η decay is, however, an old story and the results are in total disagreement if only s-wave $\pi-\pi$ interactions are considered.[13] Perhaps a more complicated model with an effective σ (at 720 MeV) ρ and f meson (origin of quadratic terms?) poles is capable of yielding a satisfactory answer. Details of parameter fitting in such a model, however, should await better statistics.

We shall conclude that the most conclusive evidence for the quadratic terms will become available when the energy spectra of pions in the neutral 3π mode is measured. Such a measurement may resolve part of the ambiguity in the calculation of η decay.

REFERENCES

1. S. L. Adler, *Phys. Rev. Letters* **14**: 1051 (1965).
2. W. I. Weissberger, *Phys. Rev. Letters* **14**: 1047 (1965).
3. C. G. Callan and S. B. Treiman, *Phys. Rev. Letters* **16**: 153 (1966).
4. Y. Hara and Y. Nambu, *Phys. Rev. Letters* **16**: 875 (1966); D.K. Elias and J. C. Taylors, *Nuove Cimento* **44**: 518 (1966); S. K. Bose and S. N. Biswas *Phys. Rev. Letters* **16**: 330 (1966): H. D. I. Aberbanel, *Phys. Rev.* **153**: 154 (1967).
5. R. Ramachandran, *Nuovo Cimento* **47A**: 669 (1967); S. K. Bose and A. M. Zimerman, Ibid **43A**: 1165 (1966); R. Graham, S. Pakvasa, and L. O'Rafeaataigh, Ibid **48A**: 830 (1967).
6. S. Weinberg, *Phys. Rev. Letters* **17**: 616 (1966); N. N. Khuri, *Phys. Rev.* **153**: 1477 (1967).
7, Columbia Berkeley-Purdue-Wisconsin-Yale Collaborators, *Phys. Rev.* **149**: 1044 (1966).
8. D. G. Sutterland, *Phys. Letters* **23**: 384 (1966).
9. W. Bardeen, L. S. Brown, B. W. Lee and H. T. Nieh, *Phys. Rev. Letters* **18**: 1170 (1967).
10. J. A. Cronin, *Phys. Rev.* **161**: 1483 (1967); (see also) S. Weinberg, *Phys. Rev. Letters* **18**: 188 (1967); J. Schwinger, *Phys. Letters* **24B**: 473 (1967); *Phys. Rev. Letters* **18**: 923 (1967); **19**: 1154 (1967); and **19**: 1501 (1967).
11. F. Crawford and L. R. Price, *Phys. Rev.* **167**: 1339 (1968); also Aditya Kumar and R. Ramachadran, T.I.F.R. preprint (unpublished).
12. S. Baltay *et al.*, report on preliminary data at the International Theoretical Physics Conference on particles and Fields, Rochester, September, 1967, (unpublished) give a value of 1.55 ± 0.25; S. Buniatov *et al.*, *Phys. Letters* **25B**: 560 (1967) give $R = 1.38 \pm 0.15$. C. Baglin *et al.*, preprint (presented at the APS Spring meeting, Washington, 1967) report 1.3 ± 0.4.
13. W. A. Dunn and R. Ramachandran, *Phys. Rev.* **153**: 1558 (1967) Similar result for τ and τ' decays was first noted by N. Khuri and S. B. Treiman, *Phys. Rev.* **119**: 1115 (1960).

Can $Y_0^*(1405)$ be a Bound State of Three Quarks?

G. Rajasekaran

TATA INSTITUTE OF FUNDAMENTAL RESEARCH
Bombay, India

Two approaches have been used so far to understand the observed hadron spectrum: a) to regard any observed hadron as a composite (bound or resonant) formed by forces between other observed hadrons, and b) to consider the observed hadrons as bound states of some so-far-unobserved heavy quarks. In the former (bootstrap model) the forces are calculated from exchanges of the known hadrons between the known hadrons and then the positions and widths (or coupling constants) of the composites are predicted. In the latter (quark model) one postulates some forces between the quarks and predicts the spectrum of quark bound states. Successes have been claimed in both these approaches.

The question is the following: can one, from the experimental knowledge of the scattering amplitudes alone, say whether a particular hadron H is a composite of other observed hadrons or a bound state of heavy quarks? We point out that there does exist a simple criterion which may allow one to choose between the two possibilities; namely that the K-matrix for the relevant observed hadron channels should have a pole near the mass of the hadron H if this hadron H is a composite of heavy quarks. Of course, the S-matrix always has a pole corresponding to a particle. We argue that the K-matrix *also* should have a pole in the neighborhood if the particle is formed by quark forces.

Recently, Kim* has determined the K-matrix for $\pi\Sigma$ and $\bar{K}N$-

---

*J. K. Kim, *Phys. Rev. Letters* **19**: 1074 (1967).

channels by a phenomenological fit to the experimental data on
$\bar{K}$–N scattering. This K-matrix does not have a pole near the Y_0^*
(1405) and so we conclude that most probably Y_0^* (1405) is not a
bound state of three quarks.

Consider a two-body two-channel S-wave scattering process, a typ-
ical example being $\pi\Sigma$, $\bar{K}$–N scattering in $I = 0$ state. The thresholds
for these two channels are denoted W_1 and W_2 in Fig. 1. Given the
2×2 S-matrix describing the two-channel process, one can intro-

W plane

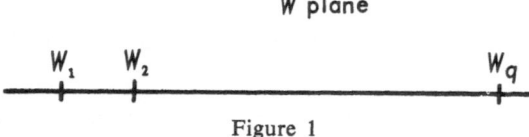

Figure 1

duce the T- and K-matrices by the relations

$$S = 1 + 2i \sqrt{k}\, T \sqrt{k} \tag{1}$$

$$T = K(1 - ikK)^{-1} \quad \text{or} \quad T^{-1} = K^{-1} - ik \tag{2}$$

where

$$k = \begin{pmatrix} k_1 & 0 \\ 0 & k_2 \end{pmatrix}$$

k_1, k_2 being the c.m. momenta in the two channels.

The elements of the T-matrix are real analytic functions of the
total energy W with branch points at W_1 and W_2. The important dis-
tinguishing property of the K-matrix is that its elements are real analytic functions of W *without the branch points W_1 and W_2*. In other
words, K is a function of k_1^2 and k_2^2 and *not* of k_1 and k_2. We are
interested in the matrix $K^{-1} = \begin{pmatrix} a & b \\ b & c \end{pmatrix}$, where a, b, and c are smooth
functions of W. A pole of K corresponds to the zero of the determi-
nant: det $K^{-1} = ac - b^2$. From (2),

$$T^{-1} = \begin{pmatrix} a - ik_1 & b \\ b & c - ik_2 \end{pmatrix}$$

so that

$$T = \frac{1}{D} \begin{pmatrix} c - ik_2 & -b \\ -b & a - ik_1 \end{pmatrix} \tag{3}$$

where

$$D = (a - ik_1)(c - ik_2) - b^2$$
$$= ac - b^2 - k_1 k_2 - (ik_1 c + ik_2 a) \tag{4}$$

This is a very convenient expression for the T-matrix since it contains the k_1, k_2 dependence explicitly.

Let us first consider the case where the mass of the hadron H lies between the thresholds W_1 and W_2 as is the case of Y_0^* (1405) which lies between the $\pi\Sigma$ and $\bar{K}N$-thresholds. This will occur as a resonance in T_{11}. From the above,

$$T_{11} = \frac{c + |k_2|}{(a - ik_1)(c + |k_2|) - b^2} = \frac{1}{a - \dfrac{b^2}{c + |k_2|} - ik_1}$$

where we have put $k_2 = +i|k_2|$.
Identifying with the usual form for elastic scattering

$$T_{11} = \frac{e^{i\delta_1} \sin \delta_1}{k_1} = \frac{1}{k_1 \cot \delta_1 - ik_1}$$

where δ_1 is the phase shift in channel 1, we have

$$k_1 \cot \delta_1 = a - \frac{b^2}{c + |k_2|}.$$

At the above mentioned resonance, $W = W_0$ (say), $\delta = \pi/2$.

$$\therefore \left\{a - \frac{b^2}{c + |k_2|}\right\}_{W=W_0} = 0$$

or

$$(ac - b^2 + a|k_2|)_{W=W_0} = 0 \tag{5}$$

In Fig. 2, we have plotted the functions $(ac - b^2)$ and $(-a|k_2|)$ schematically. Their intersection determines the resonance position W_0. Since a, b, and c are smooth functions of W, $(ac - b^2)$ can be approximated by a straight line over the small region of interest. On the other hand, $-a|k_2|$ has the form $\sim \sqrt{W_2 - W}$ near the threshold W_2. There are two distinct possibilities for $(ac - b^2)$ as illustrated in Fig. 2a and b: (a) it has a large slope and so passes through zero in the neighborhood, and (b) it has a small slope and so stays constant over the region of interest.

Let us now come back to the quark model according to which the hadron H is a composite of a fixed number of heavy quarks. Let the far-away threshold W_q in Fig. 1 denote the threshold for this quark channel. In the case of Y_0^* (1405), this will be the threshold of the three-quark channel. The details of quark model are however irrelevant for our argument. The crucial point is that since H is formed by the strong attractive forces in some channel with high threshold, the existence of H cannot depend on its relative position with respect to

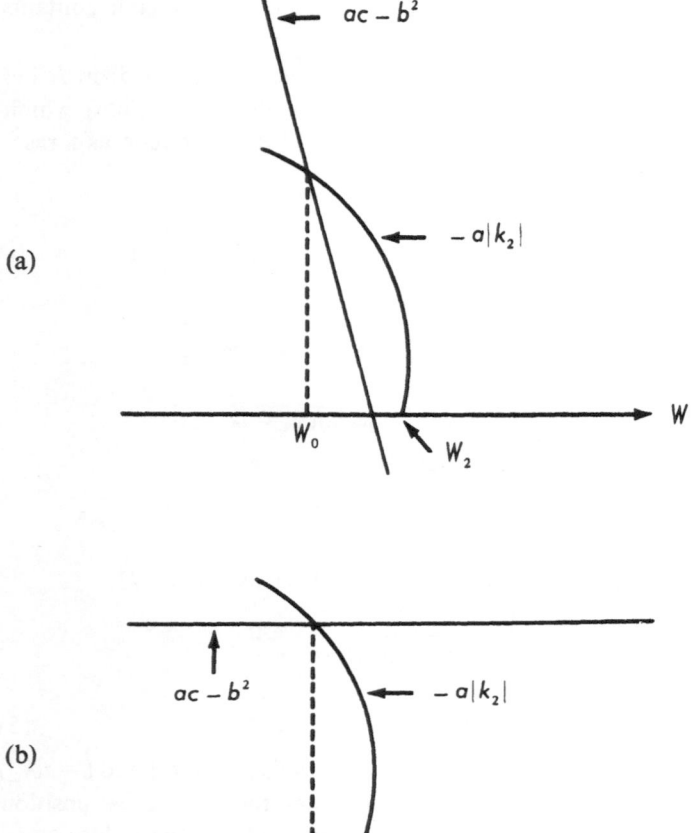

Figure 2

the meson–baryon thresholds W_1 and W_2. For instance, if we move
the threshold W_2 to the left, the value of W_0 should be relatively
unaffected and so $W_2 - W_0$ should become arbitrarily small. This is
in fact the case for case (a). For small perturbations in threshold in
which we are interested, the most sensitive dependence on the thresh-
old W_2 comes from $|k_2|$, and $(ac - b^2)$ and a can be left unchanged.
So, as threshold W_2 moves to the left, the curve $-a|k_2|$ will move
along with it whereas the line $(ac - b^2)$ stays put. The point of intersection W_0 moves only a little because of the steepness of the line
$(ac - b^2)$ and $W_2 - W_0$ approaches zero.

On the other hand, in the case of (b), one can easily see that as W_2 moves to the left, W_0 also moves to the left by about the same extent and, hence, the difference $W_2 - W_0$ remains approximately constant. One cannot decrease $W_2 - W_0$ arbitrarily in this case and so case (b) is not consistent with the expectation from quark model. In this case, H is intimately tied up with the channel 2; it is really a bound state formed by strong attractive forces in channel 2. This is the "virtual bound state resonance" of Dalitz.*

In fact, one can go further. If H is really a bound state of three quarks, then it should be possible to move the threshold W_2 to the left of W_0 and the resonance should continue to exist. Now, it will be a two-channel resonance and at the resonance $W = W_0$, one of the eigenphase-shifts passes through $\pi/2$. The connection between eigen-phase-shifts δ_α, δ_β and the K-matrix is simply

$$\frac{1}{\sqrt{k}} K^{-1} \frac{1}{\sqrt{k}} = U \begin{pmatrix} \cot \delta_\alpha & 0 \\ 0 & \cot \delta_\beta \end{pmatrix} U^{-1}$$

where U is the unitary transformation which diagonalizes $(1/\sqrt{k}) K^{-1} (1/\sqrt{k})$. (This is just the generalization of the single-channel definition: $K^{-1} = k \cot \delta$). So, at the resonance $W = W_0$,

$$\det K^{-1} = k_1 k_2 \cot \delta_\alpha \cot \delta_\beta = 0$$

Thus, if the possibility of H occuring above W_2 is not to be ruled out, i.e., if H is really formed by forces in the high-threshold channel, then $(ac - b^2)$ should necessarily pass through zero in the neighborhood.

Let us now fix our attention on Y_0^* (1405) which, as already mentioned, is connected to the $I = 0$ S-wave $\pi \Sigma$, $\bar{K}N$-channels and ask whether Y_0^* (1405) can be understood as a bound state of three quarks. We need the K-matrix for the $\pi \Sigma$, $\bar{K}N$ channels. Kim† has determined this K-matrix by a phenomenological analysis of $\bar{K}N$-scattering data and his result is (in units of F^{-1}):

$$K^{-1} = \begin{pmatrix} \pi \Sigma & \bar{K}N \\ 2.04 & -1.11 \\ -1.11 & 0 \end{pmatrix} + \begin{pmatrix} -0.45 & 0 \\ 0 & +0.27 \end{pmatrix} \begin{pmatrix} k_\Sigma^2 - k_\Sigma^2(0) & 0 \\ 0 & k_K^2 \end{pmatrix}$$

where $k_\Sigma(0)$ is the momentum in $\pi \Sigma$ channel at $\bar{K}N$-threshold.

---

* R. H. Dalitz, *Strange Particles and Strong Interactions*, Oxford University Press, 1962.

† J. K. Kim, *Phys. Rev. Letters* **19**: 1074 (1967).

$(k_\Sigma(0) = .942F^{-1})$. This is the multichannel effective range expansion of Ross and Shaw and it is found to give a very good fit to the scattering data for k_K (lab) $= 0$ to 550 MeV/c which corresponds to $W = 1435$ to 1585 MeV. Also this K-matrix predicts the position [by our equation (5)] and width [by an equation which can be obtained by differentiating (5)] of Y_0^* (1405) which are in good agreement with experimental values and so this K is valid even below the threshold as is to be expected. One can easily verify that this K-matrix does not have a pole anywhere in the neighborhood. We have plotted det K^{-1} Fig. 3 for $W \sim 1300$ to 1600 MeV over which Kim's K^{-1} can be expected to be valid and it does not pass through zero.

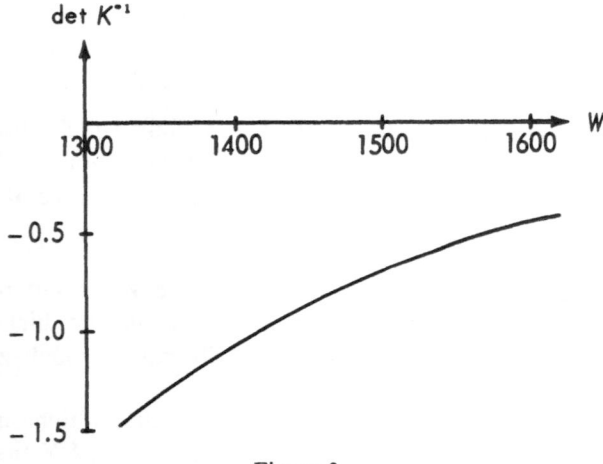

Figure 3

So we conclude that Y_0^* (1405) cannot be understood as a bound state of three heavy quarks.

The fact that Y_0^* (1405) corresponds to a "virtual bound state resonance" and that the K-matrix does not have a pole in the neighborhood is actually well known. The purpose of this note is just to point out the relevance of this fact to the quark model.

Our criterion for the validity of the quark picture, namely, that K should have a pole, works only in one direction; its converse need not be true. We have argued that if K does not have a pole in some case then the quark picture is suspect in that case. On the other hand, if K has a pole in some case, it does not necessarily follow that quark picture is correct in that case. This can be seen by the following argument.

It is known in the single-channel case that a purely attractive interaction cannot produce a S-wave resonance. A purely attractive interaction can produce a bound state if the interaction is strong enough; as the strength is reduced, the bound state moves to the right (in Fig. 1) and disappears when the threshold is reached. There is no state to the right of of the threshold. Correspondingly, $1/k \cot \delta$ does not have a pole in the neighborhood (Example: deuteron). The physical reason is that a repulsive barrier has to be present in order to create a resonance; with a repulsive barrier the composite state continues to exist even above the threshold although it is now unstable since it can leak through (see Fig. 4). A similar situation occurs in

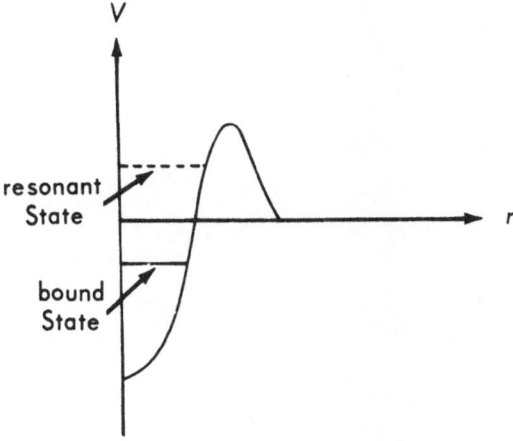

Figure 4

the multichannel case; with purely attractive interactions a S-wave resonance above the higher threshold cannot exist. (However, a resonance below the higher threshold can exist since it is just a bound state in the higher channel.) Correspondingly, the K-matrix does not have a pole. But, the interaction in the low threshold channels may really have a potential barrier in which case, as pointed out above, the state H can continue to exist even above the threshold and the K-matrix will generally have a pole. In particular, it is clear that for $l \geq 1$, the presence of the centrifugal barrier generally leads to a K-pole and, hence, one cannot say anything about the validity of the quark picture for resonances with $l \geq 1$.

The Zero Temperature Heisenberg Ferromagnet as a Field Theory

Daniel A. Dubin

IMPERIAL COLLEGE*
London, England

THE C^*-ALGEBRA OF THE SPIN 1/2 FERROMAGNET†

Physical Introduction

We envision an infinite lattice in N dimensions whose points are isomorphic to $\mathscr{Z}^{\otimes N}$. At each lattice site there is a degree of freedom known as spin. The spin can take two values which we call $\pm\frac{1}{2}\hbar$. Hereafter $\hbar = 1$, the natural units. A state of the system is given by specifying the spin ($\pm\frac{1}{2}$) at each lattice point together with an amplitude there, i.e., a real number at each site. The idealized interaction is given through a generally unbounded operator, the Hamiltonian, H, which is, in the algebraic framework, the generator of the time evolution automorphism. This simplified picture, together with a particular form for H, due originally to Heisenberg, is an idealization of the zero temperature ferromagnet where the crystal ions are frozen in place.

In formulating the mathematics of the ferromagnet we are primarily interested in considering it as a model to test the axioms of the algebraic approach, i.e., the algebra of local observables, to field theory.

---

*Visiting Senior Research Fellow; work partially supported by a USAFOSR grant; present address:Center for Theoretical Studies, University of Miami, Coral Gables, Fla.

†R. F. Streater, *Comm. Math. Phys.* (1967).

We shall use the C^*-algebra approach here, rather than the earlier W^*-algebra approach. We start from a particular representation of the algebra on a Hilbert space, which we call the product representation. Next we show that there does exist an algebra of local observables and proceed to examine the axioms usually postulated, on this model. After this we introduce a second representation, the drone representation, and show that (a) the product and drone representations are isomorphic, and (b) the abstract algebras arising from these representations are isomorphic.

The Product Representation

We shall take the lattice isomorphic to $\mathscr{L}^{\otimes 3}$ in physical situations, but for brevity we consider only $\mathscr{L}$ until further notice. With each site i, let us associate the two dimensional Hilbert space $\mathscr{H}_i$, copies of $(\mathscr{H}, \dim \mathscr{H} = 2)$. $\mathscr{H}_i$ is spanned by $|\pm>_i$, i.e., closure $\{\alpha^i|+>_i + \beta^i|->_i\} = \mathscr{H}_i$. $\mathscr{H}_i$ is equipped with a scalar product $<, >_i \colon \mathscr{H}_i \times \mathscr{H}_i \longrightarrow R$ for which $|\pm>_i$ are orthonormal. The algebra of all bounded operators on $\mathscr{H}_i$, $\mathscr{B}(\mathscr{H}_i)$ is the C^*-algebra generated by the three Pauli matrices $\sigma^{(1,2,3)}_i$; note that $|\pm>_i$ are eigenvectors of $\sigma^{(3)}_i$ with eigenvalues $(\pm 1/2)$. Also note that the Pauli matrices are matrix generators of the Lie algebra SU_2.

Let $\{i_1, i_2, \ldots, i_k\} = K \subset Z$. With every $K \subset Z$ we associate $\mathscr{H}_k = \mathscr{H}_{i_1} \otimes \cdots \otimes \mathscr{H}_{i_k}$. Clearly, $\dim \mathscr{H}_k = 2^k$, and $\mathscr{H}_k$ is spanned by $|\pm>_{i_1} \otimes \cdots \otimes |\pm>_{i_k}$. For every $i \in K$ define

$$\hat{j}^{(1,2,3)}_i = 1 \otimes \cdots \otimes \sigma^{(1,2,3)} \otimes \cdots \otimes 1 \tag{1}$$

where σ occurs at the ith place and there are k places in all. Let K, equipped with the discrete topology, be written $\mathscr{K}$. The C^*-algebra generated by $\{\hat{j}^{(\alpha)}_i \colon i \in K\}$ is written $\mathscr{A}(\mathscr{K})$ the algebra of observables associated with $\mathscr{K} \subset \mathscr{L}$. Clearly $\mathscr{A}(\mathscr{K}) = \mathscr{B}(\mathscr{H}_k)$.

We define an injection map i_{QK} for $Q \subset K$ so as to be able to define the inductive limit of the $\mathscr{A}(\mathscr{K})$. Write $\mathscr{H}_K = \mathscr{H}_Q \otimes \mathscr{H}_{K/Q}$. Then identify

$$A \in \mathscr{A}(\mathscr{Q}) \longleftrightarrow A \otimes 1_{K/Q} \in \mathscr{A}(\mathscr{K}) \tag{2}$$

This defines the injective map

$$i_{QK} \colon \mathscr{A}(\mathscr{Q}) \to \mathscr{A}(\mathscr{K}) \tag{3}$$

Following Takeda's results we consider the inductive limit of all the algebras $\mathscr{A}(\mathscr{K})$. The resulting algebra is equipped with the equivalence relation that follows. Let $A, B \in \; \uparrow \mathscr{A}(\mathscr{K})$. Let $L \subset Z$ be finite and such that $Q \subset L, K \subset L$ and

$$i_{QL}(A) = i_{KL}(B)$$

Then $A \sim B$ in $\uparrow \mathscr{A}(\mathscr{K})$. We write

$$\uparrow \mathscr{A}(\mathscr{K}) = \mathscr{A}(\mathscr{L}) \tag{4}$$

and call $\mathscr{A}(\mathscr{L})$ the abstract C^*-algebra of the ferromagnetic lattice. $\mathscr{A}(\mathscr{L})$ is the norm closure of the union of the local subalgebras $\mathscr{A}(\mathscr{K})$, $K \subset Z, K$ finite.

Let us note that for fixed $K, i, j \in K$,

$$[\hat{J}_i^{(\alpha)}, \hat{J}_j^{(\beta)}] = i \sum_{\gamma=1}^{3} \epsilon^{\alpha\beta\gamma} \hat{J}_i^{(\gamma)} \delta_{ij} \qquad \alpha, \beta = 1, 2, 3 \tag{5}$$

These relations will be called the commutation relations.

The Concrete C^*-algebra

We shall employ a certain representation of the algebra for the following reason. We define the Hamiltonian formally by

$$H = \sum_{i,j} f_{ij} \sum_{\alpha=1}^{3} \hat{J}_i^{(\alpha)} \hat{J}_j^{(\alpha)} \qquad i, j \in Z \tag{6}$$

We consider H to be a derivation of $\bigcup_K \mathscr{A}(\mathscr{K})$, acting through

$$H: A \to [H, A] \tag{7}$$

In general, the expression for H is not bounded. Firstly, we assume $f_{ij} = 0$ almost all, $i, j; f_{ij} \neq 0$ at neighbors. Secondly, H is not a continuous derivation as then it could be extended to a derivation of $\mathscr{A}(\mathscr{L})$ which was bounded in all representations. We therefore introduce the concrete C^*-algebra on which H is self-adjoint.

We start from the infinite tensor product space

$$\mathscr{H} = \bigotimes_{i \in Z}^{c} \mathscr{H}_i$$

whose elements are classes of sequences; the class structure is the equivalence relation $\psi \sim \varphi$ if $\prod_i (\psi_i, \varphi_i)$ converges.

We can now define an irreducible, faithful representation of $\mathscr{A}(\mathscr{L})$, showing that at least one such exists. Let $\mathscr{C}_{K'}$ be the equivalence class of the sequence $\{\psi_i\}_{i \in Z/K}$ obtained from ψ by omitting its projection onto $\mathscr{L}_K$. Then write

$$\mathscr{L} = \mathscr{L}_k \bigotimes_{i \in Z/k}^{\mathscr{C}_{K'}} \mathscr{L}_i \tag{8}$$

We define the norm preserving map from $\mathscr{A}(\mathscr{K}) \to \mathscr{B}(\mathscr{L})$ by

$$A \to A \otimes 1_{\mathscr{C}K'} = \bar{A} \qquad (9)$$

The C^*-algebra generated by the $\bar{A}$ defines such a faithful irreducible representation.

The concrete C^*-algebra of the ferromagnet is the incomplete product representation where equivalence is modulo the vector

$$\hat{\Omega} = \{1+>_i\} \qquad (10)$$

the ground state.

Properties of the Model

We shall list some properties of the model, omitting the proofs.

1. Let $g \in SU_2$. Representations obtained from the functionals, via the G–S construction,

$$A \to (\hat{\Omega}, A_g\hat{\Omega}) \qquad (11)$$

are all *inequivalent representations* of $\mathscr{A}(\mathscr{L})$.

2. A *complete orthonormal basis* for $\mathscr{H}^{C(\Omega)} \equiv \mathscr{H}_\Omega$ is given by the set $\{\psi_K\}$,

$$\psi_K = \hat{J}_{i_1}^{(-)}\hat{J}_{i_2}^{(-)} \cdots \hat{J}_{i_k}^{(-)}\hat{\Omega} \qquad (12)$$

$$\hat{J}_i^{(\pm)} = \hat{J}_i^{(1)} \pm i\hat{J}_i^{(2)} \qquad (13)$$

We consider $A \in D_0 = \bigcup_K \mathscr{A}(\mathscr{K})\hat{\Omega}$, $\bar{D}_0 = \mathscr{H}_\Omega$ then we redefine H by

$$H_R A\hat{\Omega} \leftarrow [H, A]\hat{\Omega} \qquad (14)$$

This *renormalization* substracts an inessential infinite additive constant. Note that if $A\hat{\Omega} = 0$, $H_R A\hat{\Omega} = 0$. Hereafter we drop the subscript R. H so defined is positive.

3. Let $\mathscr{H}^{(N)}$ be the subspace of $\mathscr{H}_\Omega$ where exactly N lattice sites have spins -1/2 (i.e., are associated with $|->$). Then $\mathscr{H}^{(N)}$ is invariant under $u(t) = e^{iHt}$. Note $\mathscr{H} = \bigoplus_1^\infty \mathscr{H}^{(N)}$.

4. The *time displacements* $A \to A_t = u(t)Au^{-1}(t)$ form a *norm continuous group of automorphisms* of $\mathscr{A}(\mathscr{L})$. This follows as (i) $\mathscr{A}(\mathscr{L})$ is countably generated, and (ii) H is self-adjoint in one representation of $\mathscr{A}(\mathscr{L})$.

5. $\mathscr{A}(\mathscr{L})$ is *asymtotically Abelian*, that is,

$$\||[A, B_a]\|| \to 0, (\vec{a} \cdot \vec{a}) \to \infty \tag{15}$$

Here $B \to B_a$ is the space translation automorphism.

6. *The Wightman functions.* Let $x_j = (t_j, \vec{n}_j \in \mathscr{L}^{\otimes 3})$ label a lattice point at a time. Then

$$W^m(x_1, \alpha_1, \ldots, x_m, \alpha_m) = (\hat{\Omega}, \hat{J}_{n_1}^{(\alpha_1)}(t_1) \cdots \hat{J}_{n_m}^{(\alpha_m)}(t_m) \hat{\Omega}) \tag{16}$$

are the Wightman functions. They are permutation invariant at equal times, satisfy the positive definite conditions (associated with the norm in $\mathscr{H}_\Omega$). The W^m have analytic continuations into the $(t_j - t_{j+1})$ product upper half planes, and the reconstruction theorem holds. There is a *strong cluster property* at equal times

$$(\hat{\Omega}, \hat{J}_1 \cdots \hat{J}_l u(\vec{n}) \hat{J}_{l+1} \cdots \hat{J}_m \hat{\Omega}) = (\hat{\Omega}, \hat{J}_1 \cdots \hat{J}_l \hat{\Omega})(\hat{\Omega}, \hat{J}_{l+1} \cdots \hat{J}_m \hat{\Omega}) \tag{17}$$

for $\vec{n} \to \infty$. As in the usual Wightman theory, we can then show that $\hat{\Omega}$ is the only vector invariant under space translations. When H is translation invariant there is a strong cluster property for arbitrary times.

7. We define states with several infinite regions, called *magnetic domains*, each with a different direction of spin. An example is the state with half spin up, half spin down. These states lead to inequivalent representations of $\mathscr{A}(\mathscr{L})$. If, and only if, the equivalence classes $\mathscr{C}(\psi)$ and $\mathscr{C}(\varphi)$ are weakly equivalent will the representations on $\mathscr{H}^{C(\psi)}$ be unitarily equivalent to the one on $\mathscr{H}^{C(\phi)}$.

Fock Space Formulation*

Let $i, j \in K$ and ψ_i, d_j be dynamically independent free spinor fields. That is, they generate a *-algebra following from

$$\{\psi_i, \psi_j^*\} = \{d_i, d_j^*\} = \delta_{ij} \tag{18}$$

others vanish.

Let us form the representation space following from the cyclic vector Ω having the Fock property

$$\psi_i \Omega = 0 \qquad d_j \Omega = 0 \tag{19}$$

Then let

---

*D. A. Dubin, *Fock Space Formulation of the Ferromagnet*, ICTP/67/36 (1967).

$$\prod_{i=1}^{k} (\psi_i^*)^{n_1(i)} \prod_{j=1}^{k} (d_j^*)^{n_2(j)} \Omega = \Phi(n_1, n_2) \tag{20}$$

Here $n_{(1,2)}(i) = 0, 1$. Those Φ for which

$$\sum_{i=1}^{k} [n_1(i) + n_2(i)] = n \tag{21}$$

is some fixed integer $\leq k$ span the space $h_n(k)$. Let

$$h(k) = \bigoplus_{n=0}^{2k} h_n(k) \tag{22a}$$

with

$$\Phi(0, 0) = \Omega \tag{22b}$$

$$h_0(k) \simeq \mathscr{C} \tag{22c}$$

also,

$$h_n(k) = h_1(k)\textcircled{a} \cdots \textcircled{a} h_n(k) \tag{22d}$$

$h(k)$ has the inner product

$$(\Phi(n_1, n_2)|\Phi(m_1, m_2)) = \delta(n_1, m_1)\delta(n_2, m_2) \tag{23}$$

The number operators

$$N_j^{(1)} = \psi_j^* \psi_j \qquad N_l^{(2)} = d_l^* d_l \tag{24a, b}$$

are defined by

$$N_j^{(1,2)} \Phi(n_1, n_2) = n_{(1,2)} \Phi(n_1, n_2) \tag{24c}$$

They are bounded and self-adjoint.

We are interested in a special subspace of $h(k)$ whick will be related to $\mathscr{H}_K$ of the previous product formulation. From the original $*$-algebra we consider the $*$-subalgebra generated by

$$j_l^{(+)} = \psi_l(d_l + d_l^*)$$

$$j_l^{(-)} = (d_l + d_l^*)\psi_l^* \tag{25}$$

$$j_l^{(3)} = -N_l^{(1)} + \tfrac{1}{2}$$

These $j_l^{(\alpha)}$ are represented as operators on $h(k)$. Let us write $\Phi(n, n) \equiv \Phi(n)$. Those $\Phi(n)$ for which $\sum_{l=1}^{k} n(l) = n, n$ a fixed integer, span the subspace $E_n(k) \subset h(k)$. Define

$$E(k) = \bigoplus_{n=0}^{k} E_n(k) \qquad \dim E(k) = 2^k \tag{26}$$

We note that as $E(k)$ is a Hilbert space, $E(k) \simeq \mathscr{H}_K$. Next we note that the $j_l^{(\alpha)}$ generate a C^*-algebra on $E(k)$, call it $\mathscr{D}(\mathscr{K})$. It is easy to show explicitly an isometric operator $u: E(k) \rightarrow \mathscr{H}_K$ and $u^{-1}: \mathscr{H}_K \rightarrow E(k)$. From the finite dimensionality, u is unitary. Further

$$u(K)\hat{J}_l^{(\alpha)}u^{-1}(K) = j_l^{(\alpha)} \qquad l \in K \tag{27}$$

Then $\mathscr{A}(\mathscr{K}) \simeq \mathscr{D}(\mathscr{K})$. It is also simple to show the validity of the commutative diagram

$$
\begin{array}{ccc}
\mathscr{A}(\mathscr{K}) & \xrightarrow{i_{QK}} & \mathscr{A}(\mathscr{Q}) \\
u(k) \downarrow & & \downarrow u(Q) \\
\mathscr{D}(\mathscr{K}) & \xrightarrow[l_{QK}]{} & \mathscr{D}(\mathscr{Q})
\end{array}
\tag{28}
$$

and so

$$\uparrow \mathscr{A}(\mathscr{K}) \simeq \uparrow \mathscr{D}(\mathscr{K}) \tag{29a}$$

or

$$\mathscr{A}(\mathscr{L}) \simeq \mathscr{D}(\mathscr{L}) \tag{29b}$$

Continuum Limit*

It is tempting to consider the situation when $\mathscr{L}$ is replaced by $\mathscr{R}$. We show how this system is self contradictory, leading to an indefinite metric when renormalized.

Our commutation relations in the continuum drone model are

$$\{\psi^*(x), \psi(y)\} = \{d^*(x), d(y)\} = \delta(x - y) \tag{30}$$

Now

$$J^{(-)}(x) = \psi^*(x)(d(x) + d^*(x)) = (J^{(+)}(x))^* \tag{31}$$
$$J^{(3)}(x) = -\psi^*(x)\psi(x) + \tfrac{1}{2}$$

The vacuum conditions are given by the action of $J^{(+)}$ and $J^{(3)}$ on the ground state Ω. We have

$$J^{(+)}(x)\Omega = 0 \tag{32}$$
$$J^{(3)}(x)\Omega = \tfrac{1}{2}\Omega$$

---

*D. A. Dubin and R. F. Streater, *Nuovo Cimento* **50**: 154 (1967).

In the continuum case, $[J^{(-)}(x), J^{(+)}(y)]\Omega$ contains a term $\sim \delta(x - y)$ $\delta(0)\Omega$. Hence, we must renormalize. We do this by smearing the $J^{(\alpha)}$ with functions of some suitable class. Then

$$[J^{(\alpha)}(f), J^{(\beta)}(g)] = i \sum_{\gamma=1}^{3} e^{\alpha\beta\gamma} J^{(\gamma)}(fg) \tag{33}$$

$$J^{(+)}(f)\Omega = 0, \qquad J^{(3)}(f)\Omega = \frac{1}{2} \int f(x)\,dx\,\Omega \tag{34}$$

Having our space equipped with a Hermitian form,

$$(\Omega, \Omega) = 1 \tag{35a}$$

$$\|J^{(-)}(f)\Omega\|^2 = \int |f(x)|^2\,d^3x \geq 0 \tag{35b}$$

and

$$\frac{1}{2}\|J^{(-)}(f)J^{(-)}(f)\Omega\|^2 = \|J^{(-)}(f)\Omega\|^4 - \int |f(x)|^4\,d^3x \tag{35c}$$

Choosing, e.g.,

$$f_0(x) = \begin{cases} 2 & -\tfrac{1}{2} < x < +\tfrac{1}{2} \\ 0 & |x| > \tfrac{1}{2} \end{cases}$$

and rounded to avoid discontinuities,

$$\tfrac{1}{2}\|J^{(-)}(f_0)J^{(-)}(f_0)\Omega\|^2 < 0 \tag{36}$$

Hence, an indefinite metric appears. It is easy to show that generalizing the commutators, to Lie fields

$$[J^{(\alpha)}(f), J^{(\beta)}(g)] = i \sum_{\alpha=1}^{3} e^{\alpha\beta\gamma} \int d^3x\,d^3y\,d^3z\, f(x)g(y)G(x - y, y - z)J^{\gamma}(z) \tag{37}$$

gives the same difficultly. It is implicit in Araki's thesis that only if $J^{(3)}$ were a stochastic field could this difficulty be avoided. Streater uses this as an example to show that his continuous tensor product construction for Hilbert spaces is not generally positive definite.*

---

*R. F. Streater, *Current Commutation Relations and Continuous Tensor Products*, ICTP/67/20 (1967).

Electromagnetic Interaction of Particles with Spin

R.H. Good, Jr.

IOWA STATE UNIVERSITY*
Ames, Iowa

---

1. INTRODUCTION

The problem considered here is that of making a relativistic theory for a particle or antiparticle, of definite nonzero mass m and spin $s = 0, \frac{1}{2}, 1, \ldots$, in an external electromagnetic field. The effects of anomalous electric and magnetic moments are to be considered also; for spin s these have to be included up to 2^{2s} pole type. This corresponds to the physical situation of an elementary particle or nucleus falling through external fields without any dissociation or internal excitation taking place.

For spin 0 there are no anomalous moments and the particle is governed by the equation

$$(\pi_\alpha \, \pi_\alpha + m^2)\psi = 0 \tag{1}$$

where $\psi(\mathbf{x}, t)$ is a single-component wave function and

$$\pi_\alpha = -i \frac{\partial}{\partial x_\alpha} - e A_\alpha$$

Here e is the charge of the particle and $A_\alpha(\mathbf{x}, t)$ is the four-vector potential $(\mathbf{A}, i\phi)$ of the external field so that

$$F_{\alpha\beta} = \frac{\partial A_\beta}{\partial x_\alpha} - \frac{\partial A_\alpha}{\partial x_\beta}$$

---

*Institute for Atomic Research and Department of Physics. This research was in part done in the Ames Laboratory of the U.S. Atomic Energy Commission.

The field tensor is related to the fields by

$$F_{ij} = \epsilon_{ijk}B_k$$
$$F_{i4} = -F_{4i} = -iE_i$$
$$F_{44} = 0$$

(The Latin indices run from 1 to 3, the Greek from 1 to 4 with $x_4 = it$. Factors of c and $\hbar$ are omitted.) Equation (1) is due to Schrödinger; in fact, he considered it as a basis for his wave mechanics before adopting $H\psi = i(\partial\psi/\partial t)$.

For spin $\frac{1}{2}$ the problem is solved by the Dirac equation with the Pauli term included to take account of the anomalous magnetic dipole moment,

$$\left(\pi_\alpha\gamma_\alpha + \frac{1}{8}\frac{e\lambda}{m}F_{\mu\nu}\gamma_\mu\gamma_\nu - im\right)\psi = 0 \tag{2}$$

The wave function has four components and the matrices satisfy

$$\gamma_\alpha\gamma_\beta + \gamma_\beta\gamma_\alpha = 2\delta_{\alpha\beta} \tag{3}$$

The constant λ is real and this describes a particle with a g-factor of $(2 + \lambda)$. A parity nonconserving electric dipole interaction can be made by putting in an additional factor of γ_5 in the Pauli term. This can always be done in the type of formulation discussed below and only the parity conserving types of interactions will be discussed.

The spin 1 particle in external fields was originally studied by Proca[1] and Kemmer[2] using a ten-component wave function. Corben and Schwinger[3] showed how to include an anomalous magnetic dipole term in Proca's theory, and Young and Bludman[4] gave an almost complete account of an anomalous electric quadrupole effect.

For spins higher than 1, a general theory has been developed by several authors starting from the work of Fierz and Pauli.[5] Fronsdal[6] has reviewed the theory and given an especially simple formulation of it. The theory takes account of normal moments only, in the sense that the fields influence the particle only by way of the potential A_μ in the operator π_μ. Thus how to account for arbitrary moments on the particles is still an open question. Incidentally, what is considered a normal moment depends on the form of the field equations that is emphasized. For example, if some components in a set of field equations are eliminated (with corresponding increase in the order of the equations) then the definition of what are the normal moments changes.

The purpose of these two lectures is to give a report of an entirely new approach to this problem of electromagnetic interaction of particles with spin. The work was stimulated by the success of Joos,[7] Weinberg,[8] and Weaver, Hammer, and Good[9] in treating free particles of spin s with a $(o, s) \oplus (s, o)$ wave function. Progress to date is as follows: (1) A complete treatment of spin 1 particles has been made including all anomalous moment effects.[10] (2) An equation for higher spins has been found but its consequences not yet developed.[11] (3) McDonald, Pursey, and Hammer[12] have given interesting equations for particles of higher spin in interaction with electromagnetic fields and dressed with neutrino fields. The spin 1 results are developed in some detail below and then a report of the other work is given, omitting most of the proofs.

2. COVARIANTLY DEFINED MATRICES FOR SPIN 1

Covariantly defined matrices were first discussed by Barut, Muzinich, and Williams.[13] The application to spin 1 was developed by Weinberg[8] and by Sankaranarayanan and Good.[14]

It is easy to see the applicability of these matrices by starting with the case of spin $\frac{1}{2}$ and the two-by-two unimodular representation of the Lorentz group. Let a Lorentz transformation be written as

$$x'_\alpha = a_{\alpha\beta} x_\beta \qquad a_{\alpha\beta} a_{\alpha\gamma} = \delta_{\beta\gamma}$$

where a_{ij}, ia_{i4}, ia_{4i}, and a_{44} are real. Consider matrices such that

$$A^+ \sigma_\alpha A = a_{\alpha\beta} \sigma_\beta \tag{4}$$

where σ_i are the Pauli matrices and σ_4 is i. It is known [15] that A can be written in the form

$$A = e^{i\tau \cdot s} \tag{5}$$

where $s = \frac{1}{2}\sigma$ and τ are three complex numbers, corresponding to the six real parameters needed to specify a Lorentz transformation continuous with the identity. For present purposes it is sufficient to consider infinitesimal Lorentz transformations in which case it is easy to verify equation (5) and identify the parameters. Infinitesimally,

$$a_{\alpha\beta} = \delta_{\alpha\beta} + b_{\alpha\beta} \tag{6}$$

where $b_{\alpha\beta} = -b_{\beta\alpha}$ and b_{ij}, ib_{i4} are real. If

$$\tau_i = b_{i4} + \tfrac{1}{2}\epsilon_{ijk} b_{jk} \tag{7}$$

then equation (5) gives

$$A = 1 + ib_{i4} s_i + \frac{i}{2} \epsilon_{ijk} s_i b_{jk}$$

$$A^+ = 1 + ib_{i4} s_i - \frac{i}{2} \epsilon_{ijk} s_i b_{jk}$$

so that

$$A^+ \sigma_\alpha A = \sigma_\alpha + \frac{i}{2} b_{i4} [\sigma_i, \sigma_\alpha]_+ - \frac{i}{4} \epsilon_{ijk} b_{jk} [\sigma_i, \sigma_\alpha]_-$$

This means that

$$A^+ \sigma_l A = \sigma_l + i b_{l4} + \tfrac{1}{2} \epsilon_{ijk} b_{jk} \epsilon_{ilm} \sigma_m$$
$$= \sigma_l + b_{l4} \sigma_4 + b_{lm} \sigma_m$$
$$A^+ \sigma_4 A = \sigma_4 - b_{i4} \sigma_i$$

and that

$$A^+ \sigma_\alpha A = \sigma_\alpha + b_{\alpha\beta} \sigma_\beta$$

as required.

Now consider generalizing equation (5) so that s are any spin matrices, not just spin $\frac{1}{2}$. There is a correspondence between Lorentz transformations and matrices $e^{i\tau \cdot s}$ since τ parametrizes the transformations. Furthermore, this correpondence is preserved under matrix multiplication so the $e^{i\tau \cdot s}$ form a representation of the Lorentz group for spin s as well as for spin $\frac{1}{2}$. One sees this easily as an application of Hausdorff's theorem[16] which says that in the product

$$e^A e^B = e^{A + B + 1/2[A,B] + 1/12[A,[A,B]] + 1/12[[A,B],B] + \cdots}$$

only higher-order commutators appear in the exponent. In a product like $e^{i\tau_A \cdot s} e^{i\tau_B \cdot s}$, every commutator simplifies like

$$[\tau_A \cdot s, \tau_B \cdot s] = i(\tau_A \times \tau_B) \cdot s$$

and the product reduces to $e^{i\tau_C \cdot s}$, the calculation leading to the same result for all spins. Thus, the correspondence is preserved under matrix multiplication for all spins and $e^{i\tau \cdot s}$ form a representation for all spins.

Now suppose that S_Γ are a complete set of $(2s + 1)$-square matrices so that any matrix of this size can be expressed as a linear combination of the S_Γ. Since they are a complete set $A^+ S_\Gamma A$ can be expressed in terms of them, say

$$A^+ S_\Gamma A = a_{\Gamma\Delta} S_\Delta \qquad (8)$$

Since the A are a representation of the Lorentz group, so are the $a_{\Gamma\Delta}$ matrices. This is the generalization of equation (4) to higher spins and the matrices S_Γ are said to be covariantly defined.

To do this in detail for spin 1 and see what representation occurs, one considers this complete set of 3×3 matrices;

$$S_{ij} = s_i s_j + s_j s_i - \delta_{ij} \tag{9a}$$

$$S_{i4} = S_{4i} = is_i \tag{9b}$$

$$S_{44} = -1 \tag{9c}$$

Here s_i are the spin 1 matrices and these are defined in such a way that

$$S_{\mu\nu} = S_{\nu\mu} \tag{10a}$$

$$S_{\mu\mu} = 0 \tag{10b}$$

These form a covariantly defined Lorentz tensor of the second rank in the sense that

$$A^+ S_{\mu\nu} A = a_{\mu\rho} a_{\nu\sigma} S_{\rho\sigma} \tag{11}$$

In verifying this, the only question is what representations occur and it is sufficient to just investigate it infinitesimally. The 44 component for example simplifies to

$$\left(1 + ib_{i4} s_i - \frac{i}{2} \epsilon_{ijk} s_i b_{jk}\right)(-1)\left(1 + ib_{i4} s_i + \frac{i}{2} \epsilon_{ijk} s_i b_{jk}\right)$$

$$= (\delta_{4\rho} + b_{4\rho})(\delta_{4\sigma} + b_{4\sigma}) S_{\rho\sigma}$$

or

$$-1 - 2ib_{i4} s_i = S_{44} + 2b_{4\rho} S_{\rho4}$$

which evidently holds.

The Lorentz transformation rule for a $(o, s) \oplus (s, o)$ wave function is

$$\psi'(x') = \Lambda\psi(x) \tag{12}$$

where

$$\Lambda = \begin{pmatrix} e^{i\mathbf{r}\cdot\mathbf{s}} & 0 \\ 0 & e^{i\mathbf{r}^*\cdot\mathbf{s}} \end{pmatrix} \tag{13}$$

Accordingly one defines matrices $\gamma_{\mu\nu}$ by

$$\gamma_{\mu\nu} = \begin{pmatrix} 0 & -S_{\mu\nu}^+ \\ -S_{\mu\nu} & 0 \end{pmatrix} \tag{14}$$

These are Hermitian, have the properties

$$\gamma_{\mu\nu} = \gamma_{\nu\mu} \tag{15a}$$

and

$$\gamma_{\mu\mu} = 0 \tag{15b}$$

and are covariantly defined in the sense that

$$\Lambda^{-1}\gamma_{\mu\nu}\Lambda = a_{\mu\rho}a_{\nu\sigma}\gamma_{\rho\sigma} \tag{16}$$

To verify this, one first calculates from equations (13) and (14) that

$$\Lambda^{-1}\gamma_{\mu\nu}\Lambda = \begin{pmatrix} 0 & -e^{-i\tau\cdot\mathbf{s}}S_{\mu\nu}^{+}e^{i\tau^{*}\cdot\mathbf{s}} \\ -e^{i\tau^{*}\cdot\mathbf{s}}S_{\mu\nu}e^{i\tau\cdot\mathbf{s}} & 0 \end{pmatrix}$$

From equation (11) it is known that

$$e^{-i\tau\cdot\mathbf{s}}S_{\mu\nu}e^{i\tau\cdot\mathbf{s}} = a_{\mu\rho}a_{\nu\sigma}S_{\rho\sigma} \tag{17}$$

Also if C_s is the matrix such that

$$C_s\mathbf{s}C_s^{-1} = -\mathbf{s}^{*}$$

then from equations (9a) it is found that

$$C_s S_{\mu\nu} C_s^{-1} = S_{\mu\nu}^{*}$$

and, hence, from the transpose of equations (17), that

$$e^{-i\tau\cdot\mathbf{s}}S_{\mu\nu}^{+}e^{i\tau^{*}\cdot\mathbf{s}} = a_{\mu\rho}a_{\nu\sigma}S_{\rho\sigma}^{+} \tag{18}$$

Equations (17) and (18) are just what is needed to make the expression for $\Lambda^{-1}\gamma_{\mu\nu}\Lambda$ simplify to $a_{\mu\rho}a_{\nu\sigma}\gamma_{\rho\sigma}$. With this type of wave function and with these covariantly defined matrices, one can make covariant spin 1 theories similar to the way that the spin $\frac{1}{2}$ theories are made with the Dirac wave function and the γ_μ matrices.

The analogue of the Dirac rules for these matrices is

$$[\gamma_{\mu\nu},\gamma_{\alpha\beta}]_{+} + [\gamma_{\mu\alpha},\gamma_{\nu\beta}]_{+} + [\gamma_{\mu\beta},\gamma_{\alpha\nu}]_{+} = 2(\delta_{\mu\nu}\delta_{\alpha\beta} + \delta_{\mu\alpha}\delta_{\nu\beta} + \delta_{\mu\beta}\delta_{\alpha\nu}) \tag{19}$$

By contracting indices you can derive

$$[\gamma_{\mu\nu},\gamma_{\alpha\nu}]_{+} = 6\delta_{\mu\alpha}$$

A complete set of 36, 6×6 matrices is built this way:

$$\gamma_1 = 1$$
$$\gamma_2 = \gamma_5$$
$$\gamma_{3,\mu\nu} = \gamma_{\mu\nu}$$
$$\gamma_{4,\mu\nu} = i\gamma_5\gamma_{\mu\nu}$$
$$\gamma_{5,\mu\nu} = i(\gamma_{\mu\lambda}\gamma_{\nu\lambda} - \gamma_{\nu\lambda}\gamma_{\mu\lambda})$$
$$\gamma_{6,\mu\rho,\nu\sigma} = [\gamma_{\mu\nu},\gamma_{\rho\sigma}]_{+} + 2\delta_{\mu\nu}\delta_{\rho\sigma} - [\gamma_{\mu\sigma},\gamma_{\rho\nu}]_{+} - 2\delta_{\mu\sigma}\delta_{\rho\nu} \tag{20}$$

where $\gamma_5 = \begin{pmatrix} -1 & 0 \\ 0 & 1 \end{pmatrix}$. The same types of matrices occur as in Dirac theory with in addition one more type. This is a general rule as the spin increases by $\frac{1}{2}$ and fits in with the fact that a new type of anomalous moment can occur. Symmetry properties of all these matrices are obvious except for type 6 which satisfies

$$\gamma_{6,\mu\rho,\nu\sigma} = -\gamma_{6,\rho\mu,\nu\sigma}$$

$$\gamma_{6,\mu\rho,\nu\sigma} = +\gamma_{6,\nu\sigma,\mu\rho}$$

$$\gamma_{6,\mu\rho,\nu\sigma} + \gamma_{6,\mu\sigma,\rho\nu} + \gamma_{6,\mu\nu,\sigma\rho} = 0 \tag{21}$$

In consequence of these symmetries there are just 10 of the γ_6 type linearly independent.

3. SPIN-1 WAVE EQUATION

The equation is

$$\left[\pi_\alpha \pi_\beta \gamma_{\alpha\beta} + \pi_\alpha \pi_\alpha + 2m^2 + \frac{e\lambda}{12} \gamma_{5,\alpha\beta} F_{\alpha\beta} + \frac{eq}{6m^2} \gamma_{6,\alpha\beta,\mu\nu} \frac{\partial F_{\alpha\beta}}{\partial x_\mu} \pi_\nu \right] \psi = 0 \tag{22}$$

Here λ and q are real constants that adjust the sizes of the intrinsic moments.

What is the nonrelativistic limit? In Ref. 10, a Foldy–Wouthuysen type of expansion on m^{-1} is developed and the wave function is separated into large and small components. A $(2s + 1)$-component nonrelativistic wave function Ψ is found that satisfies the equation

$$H\Psi = i \frac{\partial \Psi}{\partial t}$$

where the Hamiltonian is given by

$$H = \frac{\pi^2}{2m} + e\phi - \frac{e}{4m}(1 + \lambda)\mathbf{s}\cdot\mathbf{B}$$

$$+ \frac{e}{8m^2}(1 - \lambda - 2q)\left(s_i s_j + s_j s_i - \frac{4}{3}\delta_{ij}\right)\frac{\partial E_j}{\partial x_i}$$

$$- \frac{e}{8m^2}(1 - \lambda)\mathbf{s}\cdot(\boldsymbol{\pi} \times \mathbf{E} - \mathbf{E} \times \boldsymbol{\pi}) + \frac{1}{6}\frac{e}{m^2}(1 - \lambda)\nabla\cdot\mathbf{E} \tag{23}$$

The magnetic interaction term is usually written as

$$-g\frac{e}{2m}\mathbf{s}\cdot\mathbf{B}$$

so the g-factor here is $\frac{1}{2}(1 + \lambda)$. The quadrupole interaction term is conventionally written as

$$-\frac{Qe}{4s(2s+1)}\left(s_i s_j + s_j s_i - \frac{2}{3}\delta_{ij}s^2\right)\frac{\partial E_i}{\partial x_j}$$

Thus here the quadrupole moment Q is

$$Q = \frac{-1 + \lambda + 2q}{2m^2}$$

How is this related to the other spin 1 descriptions for a free particle? With e, λ, and q all zero the equation simplifies to

$$-P_\alpha P_\beta \gamma_{\alpha\beta}\psi = (P_\alpha P_\alpha + 2m^2)\psi \tag{24}$$

where $P_\alpha = -i(\partial/\partial x_\alpha)$. By operating on equation (19) with $P_\mu P_\nu P_\alpha P_\beta$ one finds that

$$(P_\alpha P_\beta \gamma_{\alpha\beta})^2 = (P_\alpha P_\alpha)^2$$

Consequently, by operating on equation (24) with $-P_\alpha P_\beta \gamma_{\alpha\beta}$, the result

$$(P_\alpha P_\alpha)^2 \psi = (P_\alpha P_\alpha + 2m^2)^2 \psi$$

is obtained. This simplifies to the Klein–Gordon equation

$$P_\alpha P_\alpha \psi = -m^2 \psi \tag{25}$$

However, one can now substitute this back into equation (24) to obtain Weinberg's equation

$$P_\alpha P_\beta \gamma_{\alpha\beta} \psi = -m^2 \psi \tag{26}$$

It is seen that equation (24) alone is equivalent to equations (25) and (26) together. This shows to some extent why the basic equation for spin 1 has the form of equation (22). If one starts from a formulation with auxiliary conditions, say equations (25) and (26), and replaces P_α by π_α to include the minimal electromagnetic interaction, the resulting pair of equations would be in contradiction. However, equation (24) stands alone and there P_α can be replaced by π_α without difficulty. To get to the Hamiltonian form from equations (25) and (26) one introduces a new wave function ψ_{WHG} by the substitution

$$\psi_{WHG} = [\tfrac{1}{2}(1 + \gamma_5) + \tfrac{1}{2}(1 - \gamma_5)\epsilon]\psi \tag{27}$$

where ϵ is the operator $E^{-1}i(\partial/\partial t)$ and E denotes $[P^2 + m^2]^{1/2}$, the positive root. Another way to write equation (25) is as

$$\epsilon^2 \psi = \psi \tag{28}$$

so by operating on equation (27) with the operator in the square brackets again, one finds

$$[\tfrac{1}{2}(1 + \gamma_5) + \tfrac{1}{2}(1 - \gamma_5)\epsilon]\psi_{WHG} = \psi \tag{29}$$

Thus, there is as much information about the system in ψ_{WHG} as there is in ψ, and either function can be used to discuss it. The equations satisfied by ψ_{WHG} are

$$P_\alpha P_\alpha \psi_{WHG} = -m^2 \psi_{WHG} \tag{30}$$

$$P_\alpha P_\beta \gamma_{\alpha\beta} \psi_{WHG} = -m^2 \epsilon \psi_{WHG} \tag{31}$$

These follow directly from equations (25) and (26) and the fact that γ_5 and $\gamma_{\alpha\beta}$ anticommute. These equations can be written in terms of three space and one time dimension:

$$(P_4^2 + P^2)\psi_{WHG} = -m^2 \psi_{WHG} \tag{32}$$

$$(P_4^2 \gamma_{44} + 2P_4 P_i \gamma_{4i} + P_i P_j \gamma_{ij})\psi_{WHG} = -\frac{m^2}{E} i \frac{\partial \psi}{\partial t} \tag{33}$$

The notation

$$\alpha = \begin{pmatrix} s & 0 \\ 0 & -s \end{pmatrix} \qquad \beta = \begin{pmatrix} 0 & 1 \\ 1 & 0 \end{pmatrix} \tag{34}$$

is used in the Hamiltonian form. These matrices are related to the $\gamma_{\mu\nu}$ matrices by

$$\gamma_{ij} = -\beta(\alpha_i \alpha_j + \alpha_j \alpha_i - \delta_{ij})$$

$$\gamma_{i4} = \gamma_{4i} = -i\beta\alpha_i \tag{35}$$

$$\gamma_{44} = \beta$$

When equation (33) is rewritten in terms of α and β and $P_4^2 \psi_{WHG}$ is replaced by $(-P^2 - m^2)\psi_{WHG}$ according to equation (32), it becomes

$$\left[-(P^2 + m^2)\beta + 2\beta\alpha \cdot \mathbf{P} i \frac{\partial}{\partial t} - 2\beta(\alpha \cdot \mathbf{P})^2 + \beta P^2 \right]\psi_{WHG}$$

$$= -\frac{m^2}{E} i \frac{\partial \psi_{WHG}}{\partial t}$$

The terms in $i(\partial/\partial t)$ combine to make

$$\left[-\frac{m^2}{E} - 2\beta\alpha \cdot \mathbf{P} \right] i \frac{\partial \psi_{WHG}}{\partial t} = [-m^2\beta - 2\beta(\alpha \cdot \mathbf{P})^2]\psi_{WHG}$$

The operator in the square brackets on the left has an inverse,

$$-\frac{E}{m^2} + \frac{2E^2 \beta\alpha \cdot \mathbf{P}}{(2E^2 - m^2)^2} + \frac{4E^3(\alpha \cdot \mathbf{P})^2}{m^2(2E^2 - m^2)^2}$$

and on multiplying through by it the equation reduces to

$$H\psi_{WHG} = i \frac{\partial \psi_{WHG}}{\partial t} \tag{36}$$

where

$$H = \frac{(2E^2 - m^2)\beta + 2E\alpha\cdot\mathbf{P} - 2(\alpha\cdot\mathbf{P})^2\beta}{2E^2 - m^2} E \qquad (37)$$

is the relativistic Hamiltonian for spin 1. In these calculations the rule $(\alpha\cdot\mathbf{P})^3 = P^2(\alpha\cdot\mathbf{P})$ is used to keep all expressions quadratic in $\alpha\cdot\mathbf{P}$. Thus, for a free spin 1 particle there is a manifestly covariant equation, equation (24) and a Hamiltonian equation, equation (36), in parallel with the Dirac situation. However, the parallel does not survive when the electromagnetic fields are included. There is the manifestly covariant equation, equation (22), but not a Hamiltonian form. Earlier, Taketani and Sakata[17] and Young and Bludman[4] relaxed the requirement that the wave function should have simple Lorentz transformation properties and obtained Hamiltonian forms of the equations of motion.

4. SPIN 3/2 AND HIGHER WAVE EQUATIONS

The spin $\frac{3}{2}$ case is sufficiently complicated that it illustrates the difficulties that come into the general case. Especially the type of solution that applies for spin $\frac{1}{2}$ or the type for spin 1 can not be extended to apply to higher spins.

The discussion of the covariantly defined matrices can be carried on. To get a complete set of $(2s + 1)$-square matrices you form all possible tensors from s, symmetric among all indices and with zero contractions, up to degree $2s$. The matrices S which are covariantly defined are built from them and form a Lorentz tensor of the $2s$-th rank symmetric among all indices and with zero contractions. The matrices $\gamma_{\alpha\beta...}$ are built from them as before and have the property

$$(\gamma_{\alpha\beta...} P_\alpha P_\beta \cdots)^2 = (P_\alpha P_\alpha)^{2s} \qquad (38)$$

The manifestly covariant description of a free particle or antiparticle is

$$\gamma_{\alpha\beta...} P_\alpha P_\beta \cdots \psi = (im)^{2s}\psi \qquad (39)$$

$$P_\alpha P_\alpha \psi = -m^2 \psi \qquad (40)$$

As a first step toward including the electromagnetic interaction terms one combines the two into a single equation:

$$\gamma_{\alpha\beta...} P_\alpha P_\beta \cdots \psi = [(P_\alpha P_\alpha)^{2s} + am^{4s-2}(P_\alpha P_\alpha + m^2)]^{1/2}\psi \qquad (41)$$

Here a is some real constant not zero. After the fields are included the effect of varying a is to vary the moments on the particle. It can be chosen to simplify some of the calculations on the equation. The right-hand side is to be considered as a power series in $(P_\alpha P_\alpha + m^2)/m^2$ and the branch of the square root to be used is the one that gives $(im)^{2s}$ as the first term in the series. To see that equation (41) is equivalent to both equations (39) and (40) for any non-zero value of a, one multiplies by $\gamma_{\alpha\beta\ldots} P_\alpha P_\beta \cdots$ and obtains

$$(P_\alpha P_\alpha)^{2s} \psi = [(P_\alpha P_\alpha)^{2s} + am^{4s-2}(P_\alpha P_\alpha + m^2)]\psi \qquad (42)$$

Thus if equation (41) holds and a is not zero, then ψ must satisfy equation (40); then on substituting back equation (41) reduces to equation (39).

An equation for the particle in the external fields is obtained by replacing P_α by π_α in equation (41) and then adding extra terms for the anomalous moments. For spin $\frac{3}{2}$ this gives

$$\gamma_{\mu\nu\rho} \pi_\mu \pi_\nu \pi_\rho \psi = [(\pi_\mu \pi_\mu)^3 + am^4(\pi_\mu \pi_\mu + m^2)]^{1/2} \psi - ep\gamma_{5,\alpha\beta} F_{\alpha\beta}$$
$$- \frac{eq}{m^2} \gamma_{6,\alpha\beta,\mu\nu} \frac{\partial F_{\alpha\beta}}{\partial x_\mu} \pi_\nu - \frac{er}{m^4} \gamma_{7,\alpha\beta \, \mu\nu,\rho\sigma} \frac{\partial^2 F_{\alpha\beta}}{\partial x_\mu \partial x_\rho} \pi_\nu \pi_\sigma$$

$$(43)$$

where p, q, r are real numbers that adjust the sizes of the moments. These spin $\frac{3}{2}$ matrices were discussed by Shay, Song, and Good.[18] Some progress in studying this equation has been made but the moments of the particles it describes are not yet known.

This formalism gives a relativistic theory but expressed as a power series about the nonrelativistic limit. The point is that the expansion parameter for the series is $(\pi_\mu \pi_\mu + m^2)/m^2$. In the nonrelativistic limit π_i is the mass times the velocity and π_4 is, for the particle, $i(m + KE)$. Thus, the expansion parameter is roughly $(v/c)^2$.

If the anomalous moment terms are disregarded the general equation is

$$\gamma_{\alpha\beta\ldots} \pi_\alpha \pi_\beta \cdots \psi = [(\pi_\alpha \pi_\alpha)^{2s} + am^{4s-2}(\pi_\alpha \pi_\alpha + m^2)]^{1/2} \psi \qquad (44)$$

This includes the spin $\frac{1}{2}$ and spin 1 theories for appropriately chosen values of a. That is, for spin $\frac{1}{2}$ and $a = -1$, the equation reads

$$\gamma_\alpha \pi_\alpha \psi = [\pi_\alpha \pi_\alpha - (\pi_\alpha \pi_\alpha + m^2)]^{1/2} \psi$$
$$= im\psi$$

Also, for spin 1 and $a = 4$, the equation is

$$\gamma_{\alpha\beta} \pi_\alpha \pi_\beta \psi = [(\pi_\alpha \pi_\alpha)^2 + 4m^2(\pi_\alpha \pi_\alpha + m^2)]^{1/2} \psi$$
$$= -(\pi_\alpha \pi_\alpha + 2m^2)\psi$$

in agreement with equation (22). For higher spins there is no choice of a that gives such a simple result.

Some completely new considerations have been made by McDonald, Pursey and Hammer. They consider free-particle equations of the form, for spin $\tfrac{3}{2}$ for example,

$$\gamma_{\mu\nu\rho}P_\mu P_\nu P_\rho \psi = imP_\mu P_\mu \psi \tag{45}$$

The point of this equation is that, if you multiply by $\gamma_{\mu\nu\rho}P_\mu P_\nu P_\rho$ it leads to

$$(P_\mu P_\mu)^3 \psi = -m^2(P_\mu P_\mu)^2 \psi$$

This means that equation (45) includes solutions for a massive particle or antiparticle satisfying

$$\gamma_{\mu\nu\rho}P_\mu P_\nu P_\rho \psi = -im^3 \psi$$
$$P_\mu P_\mu \psi = -m^2 \psi$$

and massless particles satisfying

$$\gamma_{\mu\nu\rho}P_\mu P_\nu P_\rho \psi = 0$$
$$P_\mu P_\mu \psi = 0$$

If then the electromagnetic interaction is introduced, the equation

$$\gamma_{\mu\nu\rho}\pi_\mu \pi_\nu \pi_\rho \psi = im\pi_\mu \pi_\mu \psi \tag{46}$$

describes particles and antiparticles dressed with neutrinos. It is interesting to inquire what are the properties of such objects. Equation (46) must also include solutions that go to the massless limit when the coupling is zero.

REFERENCES

1. A. Proca, *Compt. Rend*, **202**: 1490 (1936).
2. N. Kemmer, *Proc. Roy. Soc. (London)* **A173**: 91 (1939).
3. H. C. Corben and J. Schwinger, *Phys. Rev.* **58**: 953 (1940).
4. J. A. Young and S. A. Bludman, *Phys. Rev.* **131**: 2326 (1963).
5. M. Fierz and W. Pauli, *Proc. Roy. Soc.* **A173**: 211 (1939).
6. C. Fronsdal, Supp. *Nuovo Cimento* **9**: 416 (1958).
7. H. Joos, *Fortschr. Phys.* **10**: 65 (1962).
8. S. Weinberg, *Phys. Rev.* **133**: B1318 (1964).
9. D. L. Weaver, C. L. Hammer, and R. H. Good, Jr., *Phys. Rev.* **135**: B241 (1964).
10. D. Shay and R. H. Good, Jr., "Spin-one particle in an external electromagnetic field," (submitted to *Phys. Rev.*).
11. T. J. Nelson and R. H. Good, Jr., "Second quantization process for particles with spin and with internal symmetry," (submitted to *Rev. Mod. Phys.*).

12. C. L. Hammer, S. C. Mc Donald and D. L. Pursey (Preprint).
13. A. O. Barut, I. Muzinich, and D. N. Williams, *Phys. Rev.* **130**: 442 (1963); see also D. N. Williams, "Lectures in Theoretical Physics," edited by W. E. Brittin and A. O. Barut, (University of Colorado Press 1965).
14. A. Sankaranarayanan and R. H. Good, Jr., *Nuovo Cimento* **36**: 1303 (1965).
15. C. L. Hammer and R. H. Good, Jr., *Phys. Rev.* **108**: 882 (1957).
16. F. Hausdorff, *Leipzig, Ber. Ges. Wiss. Math. Phys· Kl.* **58**: 19 (1906).
17. M. Taketani and S. Sakata, *Proc. Phys. Math. Soc. Japan* **22**: 757 (1939).
18. D. Shay, H. S. Song, and R. H. Good, Jr., *Supp. Nuovo Cimento* **3**: 455 (1965).

On the Algebra of *L*-Matrices

MATSCIENCE
Madras, India

"The end of method is perspicuity"

There is a singular appropriateness in discussing the algebra of
L-matrices in this symposium in which Professor R. H. Good is one of
principal participants. For it was a seminar conducted almost twelve
years ago, in Madras, on an interesting paper of Professor Good on the
gamma matrices which started an investigation that culminated in the
present theory of *L*-matrices which include the gamma matrices within
their structure.

The fountainhead of the theory of elementary particles is the Dirac
equation which rests on four Dirac matrices $\mathscr{L}_x, \mathscr{L}_y, \mathscr{L}_z,$ and β which
are of dimension 4×4 but obey the same anticommutation relations
as the 2×2 Pauli matrices:

$$\sigma_x = \begin{pmatrix} 1 & 0 \\ 0 & 1 \end{pmatrix} \qquad \sigma_y = \begin{pmatrix} 0 & -i \\ i & 1 \end{pmatrix} \qquad \sigma_3 = \begin{pmatrix} 1 & 0 \\ 0 & -1 \end{pmatrix} \tag{1}$$

Four mutually anticommuting matrices were needed since the Dirac
Hamiltonian was postulated to be a linear combination of the four
matrices with the four quantities $\rho_x, \rho_y, \rho_z,$ and m as their coefficients,
respectively. We can obtain anticommuting matrices of higher dimen-
sions by defining the left and right direct products of Pauli matrices
with a unit matrix of arbitrary dimension as

$$I \otimes \sigma_x = S_x = \begin{bmatrix} \sigma_x & & & \\ & \sigma_x & & \\ & & \ddots & \\ & & & \sigma_x \end{bmatrix} \tag{2a}$$

$$I \otimes \sigma_y = S_y = \begin{bmatrix} \sigma_y & & & \\ & \sigma_y & & \\ & & \ddots & \\ & & & \sigma_y \end{bmatrix} \tag{2b}$$

$$I \otimes \sigma_z = S_z = \begin{bmatrix} \sigma_z & & & \\ & \sigma_z & & \\ & & \ddots & \\ & & & \sigma_z \end{bmatrix} \tag{2c}$$

$$\sigma_x \otimes I = \rho_x = \begin{bmatrix} 0 & -iIi \\ iI(-i) & 0 \end{bmatrix} \tag{3a}$$

$$\sigma_y \otimes I = \rho_y = \begin{bmatrix} 0 & -iI \\ iI & 0 \end{bmatrix} \tag{3b}$$

$$\sigma_z \otimes I = \rho_z = \begin{bmatrix} I & 0 \\ 0 & -I \end{bmatrix} \tag{3c}$$

For obvious reasons we shall call S_x, S_y, and S_z, Pauli matrices *enlarged by repetition* and ρ_x, ρ_y, and ρ_z, Pauli matrices *enlarged by dilation*. Though the dimension is increased by the direct-product operation, we obtain only sets of three anticommuting matrices and not four as required.

In the case when I is of dimension two, Dirac noticed that the set of four matrices

$$\mathscr{L}_x = \rho_x S_x \qquad \mathscr{L}_y = \rho_x S_y \qquad \mathscr{L}_z = \rho_x S_z \qquad \beta = \rho_z \tag{4}$$

will satisfy the requirements for his relativistic wave equation. The entire algebra of Dirac matrices was built out of the set of four defined above, their products, sums, and differences.

However, from an algebraic point of view the procedure of constructing the Dirac matrices from the Pauli matrices is only part of a general method of constructing higher dimensional anticommuting matrices from a primitive set of three (2×2) Pauli matrices. During the past year, a systematic study of this method has been made by the author which can be summarized as an algebra of L-matrices in the following manner.

We are aware that the only (2×2) matrix which commutes with the three Pauli matrices is a multiple of the unit matrix. *However, it is a remarkable fact that if we build "dilated" Pauli matrices ρ_x, ρ_y, ρ_z,* any matrix of the form

$$\begin{bmatrix} A & 0 \\ 0 & A \end{bmatrix} \tag{5}$$

where A is an arbitrary matrix of the same dimension as the unit matrix I in ρ_x, ρ_y, and ρ_z commutes with all the three dilated Pauli matrices. Thus, it is this commuting matrix that corresponds to the multiple of unit matrix in two dimensions.

As in the case of Pauli matrices we can form a "helicity matrix" through a linear combination of three anticommuting "dilated" Pauli matrices

$$\rho \cdot \lambda = \lambda_1 \rho_1 + \lambda_2 \rho_2 + \lambda_3 \rho_3 \tag{6}$$

From now on *we use the numerical suffix 1, 2, and 3 instead of x, y, and z for the enlarged Pauli matrices with the explicit understanding that any one of the set ρ_1, ρ_2, ρ_3 can denote ρ_x, and another ρ_y,* and the other p_z. This convention will be adopted in the ensuing discussion to simplify the notation and procedure.

For reasons which will be apparent presently we shall call the $\rho \cdot \lambda$ helicity matrix of mth order if ρ_1, ρ_2 and ρ_3 are of the dimension $2^m \times 2^m$ and denote them by $\rho_1^{(m)}, \rho_2^{(m)}, \rho_3^{(m)}$.

We can now form a set of five anticommuting matrices of dimension $2^n \times 2^n$

$$\rho_1^{(n)}, \rho_2^{(n)}, \rho_3^{(n)}, \rho_3^{(n)} \rho_1^{(n)}(n-1), \rho_3^{(n)} \rho_2^{(n)}(n-1) \tag{7}$$

where

$$\rho_i^{(n)}(n-1) = \begin{bmatrix} \rho_i^{(n-1)} & 0 \\ 0 & \rho_i^{(n-1)} \end{bmatrix} \tag{8}$$

Considering L-matrices of $(n-1)$th order we note that they commute with any one of the three matrices of order $2^{n-1} \times 2^{n-1}$

$$\begin{bmatrix} \rho_i^{(m-2)} & 0 \\ 0 & \rho_i^{(m-2)} \end{bmatrix} \tag{9}$$

and, therefore, the product of any one of the Pauli matrices of order $(n-1)$ with the above anticommutes with the other two. Thus, we arrive at seven anticommuting matrices of dimensions $2^n \times 2^n$

$$\rho_i^{(m)} \qquad (i = 1, 2)$$
$$\rho_3^{(m)} \rho_i^{(m)}(m-1) \qquad (i = 1, 2)$$
$$\rho_3^{(m)} \rho_3^{(m)}(m-1)\rho_i^{(m)}(m-2) \qquad (i = 1, 2)$$

and

$$\rho_3^{(m)}\rho_3^{(m)}(m-1)\rho_3^{(m)}(m-2) \tag{10}$$

where

$$\rho_i^{(m)}(m-2) = \begin{bmatrix} \rho_i^{(m-2)} & & & \\ & \rho_i^{(m-2)} & & \\ & & \rho_i^{(m-2)} & \\ & & & \rho_i^{(m-2)} \end{bmatrix} \tag{11}$$

These are Pauli matrices enlarged both by dilation and repetition.

This procedure can be continued until we arrive at the $(2n + 1)$ anticommuting matrices of dimension 2^n which can be conveniently arranged as $(n - 1)$ sets of two matrices and one set of three matrices

$$\rho_i^{(m)}(m) \qquad\qquad (i = 1, 2)$$

$$\rho_3^{(m)}(m)\rho_i^{(m)}(m-1) \qquad\qquad (i = 1, 2)$$

$$\rho_3^{(m)}(m)\rho_3^{(m)}(m-1)\rho_i^{(m)}(m-2) \qquad\qquad (i = 1, 2)$$

$$\rho_3^{(m)}(m)\rho_3^{(m)}(m-1)\rho_3^{(m)}(m-2)\rho_i^{(m)}(m-3) \qquad\qquad (i = 1, 2)$$

$$\vdots$$

$$\rho_3^{(m)}(m)\rho_3^{(m)}(m-1)\rho_3^{(m)}(m-2) \cdots \rho_i^{(m)}(1) \qquad\qquad (i = 1, 2)$$

and

$$\rho_3^{(m)}(m)\rho_3^{(m)}(m-1)\rho_3^{(m)}(m-2) \cdots \rho_1^{(m)}(1)$$

where

$$\rho_i^{(m)}(n-\gamma) = \begin{bmatrix} \rho_i^{(n-\gamma)} & & & \\ & \rho_i^{(n-\gamma)} & & \\ & & \ddots & \\ & & & \rho_i^{(n-\gamma)} \end{bmatrix} \tag{12}$$

where $\rho_i^{(n-\gamma)}$ is repeated 2^γ times.

This step by step procedure of obtaining $(2n + 1)$ anticommuting matrices can be expressed through a single prescription called the σ-operation by the author which can be described as follows:

Taking the primitive L-matrix of dimension 2×2

$$L_3 = \begin{bmatrix} \lambda_3 & \lambda_1 - i\lambda_2 \\ \lambda_1 + i\lambda_2 & -\lambda_3 \end{bmatrix} \tag{13}$$

which is a linear combination of the three Pauli matrices with λ_1, λ_2, and λ_3 as their respective coefficients, we obtain a matrix of dimension $(2^n \times 2^n)$ involving $(2n + 1)$ parameters by adopting the following procedure.

Replace any one of the three parameters λ_1, λ_2, and λ_3 by a matrix L_{2n-1} of dimension $(2^{n-1} \times 2^{n-1})$ and relabel the other two parameters as λ_{2n} and λ_{2n+1}, attaching unit matrices of dimension $(2^{n-1} \times 2^{n-1})$ to them. That is, we can define L_{2n+1} as

$$L_{2n+1} = \sigma(L_{2n-1}) = \begin{bmatrix} \lambda_{2n+1}I & L_{2n-1} - i\lambda_{2n}I \\ L_{2n-1} + i\lambda_{2n}I & -\lambda_{2n+1}I \end{bmatrix} \quad (14)$$

or,

$$L_{2n+1} = \begin{bmatrix} \lambda_{2n+1}I & \lambda_{2n}I - iL_{2n-1} \\ \lambda_{2n}I + iL_{2n-1} & -\lambda_{2n+1}I \end{bmatrix} \quad (15)$$

or,

$$L_{2n+1} = \begin{bmatrix} L_{2n+1} & (\lambda_{2n} - i\lambda_{2n+1})I \\ (\lambda_{2n} + i\lambda_{2n+1})I & -L_{2n-1} \end{bmatrix} \quad (16)$$

This amounts to writing L_{2n+1} as

$$L_{2n+1} = \lambda_{2n+1}\rho_3^{(n)} + \lambda_{2n}\rho_3^{(n)} + \rho_1^{(n)}\begin{bmatrix} L_{2n-1} & 0 \\ 0 & L_{2n-1} \end{bmatrix} \quad (17)$$

The first two are enlarged Pauli matrices, while the third is just the product of the third enlarged Pauli matrix and a matrix which commutes with all the three enlarged Pauli matrices. The procedure for L_{2n-1} is identical with the step by step procedure described in the definition of the σ-operation, we can replace *any one* of the parameters by L_{2n-1}. This feature is best expressed through the use of numeral suffixes for the enlarged Pauli matrices. Writing L_{2m+1} as

$$L_{2m+1} = \sum_{i=1}^{2n+1} \lambda_i \mathscr{L}_i^{2m+1} \quad (18)$$

we immediately notice that the L-matrices can be expressed as products of ρ-matrices of various orders as given in equation (10). The algebra of the L-matrices follows from the algebra of the ρ-matrices as exemplified in equation (12). So we get the product of $(2m + 1)$ L-matrices as

$$\mathscr{L}_1^{(2m+1)} \mathscr{L}_2^{(2m+1)} \cdots \mathscr{L}_{(2m+1)}^{(2m+1)} = i^m I \quad (19)$$

L_{2n+1} is a function of $2n + 1$ parameters λ_1, λ_2, and λ_{2n+1} and

$$L^2 = (\lambda_1^2 + \lambda_2^2 + \cdots + \lambda_{2n+1}^2)I = \Lambda_n^2 I \quad (20)$$

The Dirac Hamiltonian was identified to be L_5 with $\lambda_1 = p_x$, $\lambda_2 = p_y$, $\lambda_3 = p_z$, $\lambda_4 = 0$, and $\Lambda_2 = E$

$$L_5^2 = E^2 I = (p_x^2 + p_y^2 + p_z^2)I + m^2 I \quad (21)$$

where p_x, p_y, p_z are the three components of momenta, m the mass, and E the energy.

To establish closer contact with relativistic transformations we define

$$v_1 = \frac{\lambda^1}{\Lambda}\,;\, v_2 = \frac{\lambda_2}{\Lambda}\,;\, \ldots\,;\, v_{2n} = \frac{\lambda_{2n}}{\Lambda}\,;\, \lambda = \lambda_{2n+1} \qquad (22)$$

The parameters $\lambda_1, \ldots, \lambda_{2n-1}, \lambda_{2n}$ can then be expressed in terms of $\lambda, v_1, v_2, \ldots, v_{2n}$ as

$$\lambda_1 = \frac{\lambda v_1}{\sqrt{1-v^2}}\,;\, \lambda_2 = \frac{\lambda v_2}{\sqrt{1-v^2}}\,;\, \ldots\,;\, \lambda_{2n} = \frac{\lambda v_{2n}}{\sqrt{1-v^2}} \qquad (23)$$

where

$$v^2 = \frac{\Lambda^2 - \lambda^2}{\Lambda^2} + v_1^2 + \cdots + v_{2n}^2 \qquad (24)$$

REFERENCES

1. A. Ramakrishnan, "The Dirac Hamiltonian as a member of a hierarchy of matrices," *J. Math. Anal. Appl.* **20**: 9-16 (1967).
2. A. Ramakrishnan, "Helicity and energy as members of a hierachy of eigenvalues," *J. Math. Anal. Appl.* **20**: 397-401 (1967).
3. A. Ramakrishnan, "Symmetry operations on a hierarchy of matrices," *J. Math. Anal. Appl.* **21**: 39-42 (1968).
4. A. Ramakrishnan and I. V. V. Raghavacharyulu, "A note on the representations of Dirac groups," in: Symposia on Theoretical Physics and Mathematics, Vol. 8, Plenum Publishing Corp, New York, 1968, pp. 25-32.
5. A. Ramakrishnan, "On the relationship between the L-matrix hierarchy and Cartan spinnors," *J. Math. Anal. Appl.* 1967 (in press).
6. A. Ramakrishnan, R. Vasudevan, N. R. Ranganathan and P. S. Chandrasekaran, "A generalisation of the L-matrix hierarchy," *J. Math. Anal. Appl.* (in press).
7. A. Ramakrishnan, "L-matrices, quaternions and propagators," *J. Math. Anal. Appl.* (in press).
8. A. Ramakrishnan, "New perspectives on the Dirac Hamiltonian and the Feynman propagator," Proceedings of the Colorado Summer School, 1967 (in press).
9. A. Ramakrishnan, "A New form for the Feynman propagator," *J. Math. Phys. Sciences*, Vol. I, June 1967, Nos. 1 and 2 p. 57.
10. A. Ramakrishnan, "L-Matrix hierarchy and the higher dimensional Dirac Hamiltonian," I. I. T. Journal of Mathematical and Physical Sciences, Madras (in press)
11. A. Ramakrishnan, "A Hierarchy of helicity operators in L-matrix theory," I. I. T. *J. Math. Phys. Sciences*, Madras (in press),

L-Matrices and Propagators with Imaginary Parameters

ALLADI RAMAKRISHNAN

MATSCIENCE
Madras, India

---

1. INTRODUCTION

The study of L-matrices leads us naturally to the definition of the matrices

$$Q = L - \lambda I \tag{1}$$

and

$$R = \frac{1}{L - \lambda I} \tag{2}$$

where Q is a "quaternion-like" object and its reciprocal R is the resolvent which can be interpreted as the propagator associated with L under "suitable" circumstances.

The suitable circumstances relate to the interpretation of the 2_{n+1} parameters $\lambda_1, \lambda_2 \cdots$, and λ_{2n+1} which are imbedded in L such that

$$L^2 = (\lambda_1^2 + \lambda_2^2 + \cdots + \lambda_{2n+1}^2)I \tag{3}$$
$$= \Lambda_n^2 I$$

where $\pm \Lambda_n$ are the two eigenvalues of the $2^n \times 2^n$ dimensional matrix and I is a unit matrix of the same dimension as that of L. It has been emphasized by the author[1] that these parameters are either pure real or pure imaginary, which implies that the eigenvalue also is pure real or pure imaginary.

The distinction between real and imaginary parameters and the corresponding eigenvalues becomes significant when we are interested in the existence of a Fourier transform of L with respect to the parameters. To understand the consequences which follow from the introduction of imaginary quantities, we recall a very simple feature in the integration of an exponential function $e^{a\tau}$ over an infinite domain in τ, from $-\infty$ to ∞.

Case 1: If a is real and negative, then the integral is finite even when the upper limit is ∞. The lower limit should not extend to $-\infty$. Without loss of generality we can set the lower limit to be zero. The integral is

$$\int_0^\infty e^{a\tau}d\tau = -\frac{1}{a} \tag{4}$$

Case 2: When a is read positive, the lower limit can be $-\infty$ but the upper limit should not extend to ∞. We have for the integral

$$\int_{-\infty}^0 e^{a\tau}d\tau = \frac{1}{a} \tag{5}$$

Case 3: When a is pure imaginary, the integration can be performed over the entire domain. If $a = i\beta$, the integral

$$\int_{-\infty}^\infty e^{i\beta\tau}d\tau = 2\pi i\delta(\beta) \tag{6}$$

where δ is the Dirac delta function.

Case 4: When a is complex, set $a = \beta + i\gamma$. The domain of integration will depend upon whether β is positive or negative, as given in equations (4) and (5).

2. FOURIER TRANSFORMS OF THE RESOLVENT

We now define the Fourier transform of the resolvent with respect to a *partial set* of variables $\lambda_1, \ldots, \lambda_p$. For reasons of convenience, we relabel the other parameters $\lambda_{p+1}, \lambda_{p+2}, \ldots, \lambda_{2n+1}$ as m_1, $m_2, \ldots, m_{2n-p+1}$. Let us define

$$P^2 = \lambda_1^2 + \lambda_2^2 + \cdots + {}_p^2 \tag{7}$$

and

$$M^2 = m_1^2 + m_2^2 + \cdots + m_{2n-p+1}^2 \tag{8}$$

Thus, $\Lambda_n^2 = P^2 + M^2$. We call the parameters $\lambda_1, \lambda_2, \ldots, \lambda_p$ "momentum-like," since the Fourier transformation is defined with respect to them, the parameters $m_1, \ldots, m_{2n-p+1}$ are called "mass-like" since they are kept constant in the Fourier transformation, and Λ_n is "energy-like" since it is an eigenvalue. The variables $x_1, \ldots, x_p$ of the transform associated with $\lambda_1, \ldots, \lambda_p$ are called "space-like" while the variable t associated with free parameter λ is called "time-like."

To facilitate the discussion on various types of particles later in this paper, we now introduce "velocity-like" parameters

$$v_1 = \frac{\lambda_1}{\Lambda_n}; \ v_2 = \frac{\lambda_2}{\Lambda_n}; \ldots ; \ v_p = \frac{\lambda_p}{\Lambda_n} \tag{9}$$

We can express the parameters $\lambda_1, \ldots, \lambda_p$ in terms of $v_1, \ldots, v_p$ and M as

$$\lambda_1 = \frac{Mv_1}{\sqrt{1-v^2}}; \ \lambda_2 = \frac{Mv_2}{\sqrt{1-v^2}}; \ldots; \ \lambda_p = \frac{Mv_p}{\sqrt{1-v^2}} \tag{10}$$

with

$$v^2 = \frac{\Lambda_n^2 - M^2}{\Lambda_n^2} = v_1^2 + \cdots + v_p^2 \tag{11}$$

We will now classify "particles" as follows, making the postulate that the integration over the parameters is from $-\infty$ to ∞:

Case 1: We now assume that $p^2 > 0, M^2 > 0$, hence, $\Lambda_n^2 > 0$. The denominator has singularities at $\pm\Lambda_n$ on the real axis. The integral will depend upon the path of integration chosen suitably. The situation is identical to the definition of propagators corresponding to the advanced, retarded, and Feynman kernels so well known in quantum electrodynamics. The integrals corresponding to the well-known paths of integration are

$$K_F = \int_{-\infty}^{\infty} d\lambda_1 \cdots d\lambda_p d\lambda \frac{1}{2\Lambda_n} \left\{ \frac{L_{2n+1} + \Lambda_n I}{\lambda - \Lambda_n + i\epsilon} - \frac{L_{2n+1} - \Lambda_n I}{\lambda + \Lambda_n - i\epsilon} \right\}$$
$$\times e^{i}(\lambda_1 x_1 + \cdots + \lambda_p x_p) - i\lambda t \tag{12}$$

$$K_R = \int_{-\infty}^{\infty} d\lambda_1 \cdots d\lambda_p d\lambda \frac{1}{2\Lambda_n} \left\{ \frac{L_{2n+1} + \Lambda_n I}{\lambda - \Lambda_n + i\epsilon} - \frac{L_{2n+1} - \Lambda_n I}{\lambda + \Lambda_n + i\epsilon} \right\}$$
$$\times e^{i}(\lambda_1 x_1 + \cdots + \lambda_p x_p) - i\lambda t \tag{13}$$

$$K_A = \int_{-\infty}^{\infty} d\lambda_1 \cdots d\lambda_p d\lambda \frac{1}{2\Lambda_n} \left\{ \frac{L_{2n+1} + \Lambda_n I}{\lambda - \Lambda_n - i\epsilon} - \frac{L_{2n+1} - \Lambda_n I}{\lambda + \Lambda_n - i\epsilon} \right\}$$
$$\times e^{i}(\lambda_1 x_1 + \cdots + \lambda_p x_p) - i\lambda t \tag{14}$$

Integrating with respect to variable λ we obtain

$$Q_F = \int_{-\infty}^{\infty} \frac{1}{2\Lambda_n} \{\theta(t)[L_{2n+1} + \Lambda_n I]\, e^{-i\Lambda_n t} + \theta(-t)[L_{2n+1} - \Lambda_n I]\, e^{i\Lambda_n t}\}$$

$$\times\, d\lambda_1 \cdots d\lambda_p \tag{15}$$

$$Q_R = \int_{-\infty}^{\infty} \frac{1}{2\Lambda_n} \{\theta(t)[(L_{2n+1} + \Lambda_n I)e^{-i\Lambda_n t} - (L_{2n+1} - \Lambda_n I)e^{i\Lambda_n t}]\}$$

$$\times\, d\lambda_1 \cdots d\lambda_p \tag{16}$$

$$Q_A = \int_{-\infty}^{\infty} \frac{1}{2\Lambda_n} \theta(-t)[(L_{2n+1} + \Lambda_n I)e^{-i\Lambda_n t} + (L_{2n+1} - \Lambda_n I)e^{i\Lambda_n t}]$$

$$\times\, d\lambda_1 \cdots d\lambda_p \tag{17}$$

Case 2: Assume $M^2 < 0$. As $\Lambda_n^2 = (p^2 + M^2)$, Λ_n is real when $p^2 \geq M^2$ and Λ_n will be imaginary when $p^2 < M^2$. Thus, the integral splits into two parts: corresponding to $p^2 \geq M^2$ and $p^2 < M^2$. For the region, $p^2 \geq M^2$, the above expressions for advanced, retarded, and Feynman kernels will still hold good. However, when $p^2 < M^2$ is pure imaginary the exponential involves a real exponent and, hence, corresponds to a rapidly decreasing or increasing function of t.

All these features can be imbedded in the above kernels provided we require Λ_n to be pure positive real or pure negative imaginary, in the case of imaginary mass and use the above expressions for the kernels.

Case 3: Let us assume $p^2 < 0$. Now the integration with respect to space-like variables has to be split up into two halves. This division looks artificial and, hence there is a difficulty in understanding its significance. However, in the case of a "radial" distributions we can integrate in space over the semi-infinite range 0 to ∞. The physical meaning will be explained presently.

From the above discussion, it is clear that it is possible to have imaginary values for a partial set of parameters in the L-matrix provided the domains of space-like and time-like variables are divided dichotomously into positive and negative domains.*

3. PHYSICAL INTERPRETATION

To discuss the relevance of the above considerations to physical problems we now set the parameters in L_5 as follows:

---

* In such case it is called Fourier–Carleman transformation familiar in the theory of Fourier transforms.

$$\lambda_1 = p_x; \ \lambda_2 = p_y; \ \lambda_3 = p_z; \ \lambda_4 = m_1 = 0; \ \lambda_5 = m_2 = M \qquad (18)$$

With this choice we can classify particles according to the following prescription:

1. Ordinary Free Particles:

$$M^2 \geq 0 \qquad p^2 \geq 0 \qquad E^2 \geq 0$$

In this case, there exists a rest system ($p = 0$) for the particle when $M^2 > 0$. However, when $M = 0$, there is no rest system as is in the case for photons or neutrons.

2. If $M =$ imaginary, $E^2 < p^2$. Two classes can now be distinguished: (a) $E^2 > 0$, i.e., E is real and (b) $E^2 > 0$; E is imaginary. Tanaka[2] has considered this question in great detail by assuming the mass to be imaginary in the Dirac equation and postulating "superlight particles" with the velocities greater than that of light. We see from equations (12), (13), and (14) that the expression for the propagators in cofiguration space with imaginary energies involves an exponential decaying function of time. However, the integrals for these functions exist for time integration from 0 to ∞. If we consider the wave functions of a particle corresponding to this, it decays almost instantaneously, i.e., *it is "evanescent."* However, such particles will make a contribution to the momentum transform of the resolvent. This means that, though we cannot have incident or emergent particles with imaginary energy, they can still contribute to the energy denominators. These particles have imaginary velocities. Sudarshan *et al.*[3,4] have assumed that particles of imaginary mass must have real energy but the propagator formalism *necessitates the inclusion of evanescent particles of imaginary energy and velocity when considering particles of imaginary mass.*

3. $p^2 < 0; \ M^2 > 0$ with $-p^2 < M^2$. This case corresponds to bound particles. If we wish to make a space integration from 0 to ∞, we must have a radial integral.

Hence, bound particles have as much extension in space in relation to the universe as evanescent particles have existence in time in relation to eternity. Further, just as bound particles with imaginary momentum are expected to play a role in the scattering of real particles, the evanescent particles should play a similar role in the interactions of particles with imaginary mass. It is expected that these considerations will be important in postulating the possible existence of faster-than-light particles,[5] and examining the consistency of such a postulate with the Lorentz group.

ACKNOWLEDGMENT

This paper was stimulated by the lively discussion at the Matscience Symposium on "Faster-than-light particles" held at Madras on March 3, 1968 with Professor E. C. G. Sudarshan, Syracuse University, as the principal speaker.

REFERENCES

1. A. Ramakrishnan, "L-matrices, quaternions, and propagators," *J. Math Anal. Appl.* (in press).
2. Sho. Tanaka, "Theory of matter with superlight velocity," *Progr. Theor. Phys.* **24**: 171 (1960).
   E. P. Wigner, "Invariant Quantum Mechanical Equations," Sec. VII "Case of Imaginary Rest Mass," p. 76, Proceedings of the International Seminar on Theoretical Physics Trieste, I. A. E.A., Vienna, 1963.
3. O. M. P. Bilaniuk, V. K. Deshpande, and E. C. G. Sudarshan, "Meta-Relativity," *Am. J. Phys.* **30**: 718 (1962).
4. M. E. Arons and E. C. G. Sudarshan, "Lorentz invariance," "Local field theory, faster-than-light-particles," Syracuse Preprint NYO-3399-133, Su-1206-133.
5. G. Feinberg, "Possibility of faster-than-light particles," *Phys. Rev.* **159**: 1089 (1967).

A Hierarchy of Idempotent Matrices

ALLADI RAMAKRISHNAN AND R. VASUDEVAN

MATSCIENCE
Madras, India

We notice that any (3×3) antisymmetric matrix

$$
\begin{bmatrix}
0 & -\lambda_3 & \lambda_2 \\
\lambda_3 & 0 & -\lambda_1 \\
-\lambda_2 & \lambda_1 & 0
\end{bmatrix}
\tag{1}
$$

where $\lambda_1, \lambda_a, \lambda_3$ are pure real or pure imaginary parameters, has the very interesting property

$$
A^3 = (\lambda_1^2 + \lambda_2^2 + \lambda_3^2)A
\tag{2}
$$

The determinant of A is zero; hence, one of the three eigenvalues of A is zero, the other two being given by $\pm\Lambda_1$ where

$$
\Lambda_1 = +\sqrt{\lambda_1^2 + \lambda_2^2 + \lambda_3^2}
\tag{3}
$$

The eigenvectors corresponding to the eigenvalues $\pm\Lambda_1$ and 0 are

$$
\omega_+ = \begin{pmatrix} i\lambda_1\lambda_3 + \Lambda_1\lambda_2 \\ i\lambda_2\lambda_3 - \lambda_1\Lambda_1 \\ -i(\lambda_1^2 + \lambda_2^2) \end{pmatrix} \quad
\omega_- = \begin{pmatrix} i\lambda_1\lambda_3 - \Lambda_1\lambda_2 \\ i\lambda_2\lambda_3 + \lambda_1\Lambda_1 \\ -i(\lambda_1^2 + \lambda_2^2) \end{pmatrix} \quad
\omega_3 = i\lambda_3\begin{pmatrix} \lambda_1 \\ \lambda_2 \\ \lambda_3 \end{pmatrix} \tag{4}
$$

The matrix A^2 is nondiagonal. Though it is a 3×3 matrix, it has only two eigenvalues Λ_1^2 and 0. There are two independent eigenvectors which can be obtained as linear combination of the two eigenvectors of A corresponding to the eigenvalue $\pm\Lambda_1$. It is to be noted that while an eigenvector A is an eigenvector of A^2, the converse is not true since an

85

eigenvector of A^2 corresponding to Λ^2 may be a linear combination of the eigenvector of A with different eigenvalues of $\pm\Lambda$. Hence, it is not an eigenvector of A.

Writing equation (2) as

$$A_3^2 A_3 = \Lambda_1^2 A_3 \tag{5}$$

we immediately recognize that the three columns of A are just the eigenvectors of A^2 corresponding to the eigenvalue Λ_1^2. Out of these columns only two are independent since A is singular. This is as it should be, since these eigenvectors correspond only to the doubly degenerate eigenvalue Λ_1^2.

In the theory of L-matrices,[1,2] we started with the primitive (2×2) L-matrix L_3 such that

$$L_3^2 = (\lambda_1^2 + \lambda_2^2 + \lambda_3^2)I \tag{6}$$

To obtain the matrix L_{2n+1} of dimension 2^n having $(2n + 1)$ parameters, we replace any one of the three parameters in (6) by L_{2n-1} and the other two by $(\lambda_{2n+1})I$ and $(\lambda_{2n})I$, where I is a unit matrix of the same dimension as L_{2n-1}.

In the case of the A-matrix we can adopt a similar procedure if we recognize that the numerical coefficient of A on the right-hand side of equation (5) is quadratic in λ_1, λ_2, and λ_3. In the A-matrix we can replace any one of the parameters by L_{2n-1} of dimension 2^{n-1} and the other two parameters by $\lambda_{2n}I$ and $\lambda_{2n+1}I$. We thus obtain

$$A_{2n+1} = \begin{bmatrix} 0 & L_{2n-1} & \lambda_{2n+1}I \\ -L_{2n-1} & 0 & -\lambda_{2n}I \\ -\lambda_{2n+1}I & \lambda_{2n}I & 0 \end{bmatrix} \tag{7a}$$

or

$$A_{2n+1} = \begin{bmatrix} 0 & \lambda_{2n+1}I & L_{2n-1} \\ -\lambda_{2n+1}I & 0 & -\lambda_{2n}I \\ -L_{2n-1} & \lambda_{2n}I & 0 \end{bmatrix} \tag{7b}$$

or

$$A_{2n+1} = \begin{bmatrix} 0 & \lambda_{2n+1}I & \lambda_{2n}I \\ -\lambda_{2n+1}I & 0 & -L_{2n-1} \\ -\lambda_{2n} & L_{2n-1} & 0 \end{bmatrix} \tag{7c}$$

It is to be emphasized that higher dimensional A-matrices are obtained by imbedding an L-matrix and not the A-matrix.

The eigenvalues of A_{2n+1} are $\pm \Lambda_n$ and 0, where

$$\Lambda_n = \pm \sqrt{\lambda_1^2 + \lambda_2^2 + \cdots + \lambda_{2n+1}^2} = \sqrt{\Lambda_{n-1}^2 + \lambda_{2n}^2 + \lambda_{2n+1}^2} \qquad (8)$$

In the case of the L-matrices we were able to identify the *helicity matrix* to be L_3 with $\lambda_1 = p_x$; $\lambda_2 = p_y$; $\lambda_3 = p_z$ and the *Dirac Hamiltonian* with L_5 setting

$$\lambda_1 = p_x; \quad \lambda_2 = p_y; \quad \lambda_3 = p_z; \quad \lambda_4 = 0; \quad \lambda_5 = m \qquad (9)$$

The question now arises whether a similar interpretation can be attempted in the case of the A-matrices which involve one set of trichotomous eigenvalues and the rest are dichotomous sets.

In the case of A_3, there seems to be a direct connection with the polarization states of the photon. Since the operator A_2 yields only the square of energy, is it meaningful to speculate that this implies that the particle and the antiparticle of photon are the same?

In the case of A_5, if we choose L_3 to be the helicity matrix with $\lambda_1 = p_x$; $\lambda_2 = p_y$; $\lambda_3 = p_z$ and if set $\lambda_4 = 0$; $\lambda_5 = m$, are the eigenvectors of A_5 the relativistic eigenstates of an elementary particle? In this case it will still represent a particle of spin $1/2$ if L_3 is chosen to be the helicity matrix, while the meaning of the trichotomous eigenvalue is not clear.

The situation is not similar to that of the case of which is directly related to the Dirac Hamiltonian.

The difficulty arises because we are imbedding L_3 in A_5 and not an A_3 in an L-matrix. To obtain relativistic state vectors for higher spin, we must imbed A_3 in a matrix which leads to the quadratic relativistic relation. Such attempts have led some workers to make linear combinations of matrices with nonlinear coefficient parameters. These attempts have been described in detail by Professor R. H. Good in this symposium.[3]

ACKNOWLEDGMENT

It is a pleasure to thank Dr. N. R. Ranganathan for interesting discussions.

REFERENCES

1. A. Ramakrishnan, "The Dirac Hamiltonian as a member of a hierarchy of matrices," *J. Math. Anal. Appl.* **20**: 9–19 (1967).

2. A. Ramakrishnan, "On the algebra of L-matrices," in: *Symposia on Theoretical Physics and Mathematics, Vol*, 9, Plenum Publishing Corp., New York, 1968, pp. 73–78.
3. R. H. Good, "Electromagnetic interaction of particles with spin," in: *Symposia on Theoretical Physics and Mathematics, Vol.* 9, Plenum Publishing Corp., New York, 1968, pp. 59–71.

Photon Statistics and Coherence in Light Beams

R. VASUDEVAN

MATSCIENCE
Madras, India

---

1. INTRODUCTION

It was the thesis of Toynbee that great religions arise by the encounter of two great civilizations. The Graeco-Roman civilization meeting the Syrian civilization is supposed to have produced Christianity. Be that as it may, we know for certain that meeting of two disciplines of thought or even two subdisciplines in the same categories produce fascinating results. One such very fruitful encounter is the application of stochastic theory of point processes to the theory of interference in light beams. A new element was injected into this field, by the well-known Brown and Twiss experiment[1] which opened up a new chapter on "photon statistics and coherence" in optics. A semiclassical study of these coherence properties and the statistics of the ejected electrons has been carried out by Glauber, Sudarshan, and others[2] using P-representations or quasi-probability distributions. These are density matrix descriptions obtained by a knowledge of the coherent state formulation of the photon fields. Mandel[3] adopted a simpler procedure of obtaining the first few moments of the number of photoelectrons ejected in time internal $[0, T]$, by taking the correlations in intensity of the photon beam at different times. They all found that the first few moments coincided with those of the Bose distribution for the number of photoelectrons emitted by thermal light. However, a general method can be adopted to find the probability distribution of photoelectrons if one knows all orders of correlations in intensity existing in the incident beam. Assuming that a classical probability distribution of photoelectrons exists, one can always arrive at $P(n, T)$, the probability for n electrons to be ejected in time $[0, T]$ using formulas pertaining to stochastic point processes. This is possible since one can easily formulate the problem in terms of the product densities[4] (see below) of the counts occuring at various times. This formulation also leads in a straightforward manner the probability distribution that would result if two or more light beams are mixed and fall on the photo-detector. If only we know the statistical correlations in each beam, we can find the $P(n, T)$ using generalized cumulant expansions. The main purpose of this paper, in addition to surveying the main features of photostatistics and coherent beams brought to light by the recent developments in optical coherence theory, is to point out that many results deduced from a knowledge of P-representation or quasi-probability distributions can be made apparent by well-known methods of stochastic point processes. All that one needs is to be in possession of different orders of correlations that characterize the beam.

2. STOCHASTIC POINT PROCESSES METHODS

To use the powerful methods of stochastic point processes, it is necessary to deal with product densities introduced by Ramakrishan and used widely in various situations by Ramakrishnan and others.[5] This relates to the stochastic problem of the distribution of a discrete number of entities in a continuous infinity of states. A case in a point is the study of the electrons in a cascade shower in certain continuous energy ranges. For this subject, we are concerned with the statistical properties of the ejected number of photoelectrons in a given interval of time which again is a continuous parameter. The central quantity of interest in this situation is $dN(t)$, the number of electrons ejected in time t and $t + dt$. We assume that the probability that there is one electron between t and $t + dt$ is proportional to dt, while the probability that there is more than one is of order smaller than dt. Hence, the mean number in dt is

$$\epsilon\{dN(t)\} = f_1(t)\, dt$$

while

$$\epsilon\{dN(t)^r\} = f_1(t)\, dt \tag{1}$$

where $f_1(t)$ is called the product density of the first order. Product densities of higher order express all the correlations of the stochastic variable $dN(t)$ existing at various times.

$$\epsilon\{dN(t_1)\,dN(t_2)\} = f_2(t_1t_2)\,dt_1\,dt_2$$
$$\epsilon\{dN(t_1)\,dN(t_2)\cdots dN(t_\gamma)\}$$
$$= f_\gamma(t_1t_2\cdots t_\gamma)\,dt_1\,dt_2\cdots dt_\gamma \quad (2)$$

where f_γ are the higher-order product densities. The mean number of entities in a given range of the parameter t, is given by

$$\epsilon\{N(b) - N(a)\} = \epsilon\int_a^b dN(t) = \int_a^b f_1(t)\,dt \quad (3)$$

Similarly, the mean-square number of entities in the range a to b of t is

$$\epsilon\{[N(b) - N(a)]^2\} = \iint \epsilon\{dN(t_1)\,dN(t_2)\}$$
$$= \int_a^b f_1(t)\,dt + \int_a^b\int_a^b f_1(t_1, t_2)\,dt_1\,dt_2 \quad (4)$$

Equation (4) brings out the singular behavior of the random variables $dN(t_1)\,dN(t_2)$ when t_1 and t_2 coalesce. Ramakrishnan[4] has proved a very useful result for the calculation of the γth moment of the number of entities in the desired range. It runs as

$$\epsilon\{[N(b) - N(a)]^\gamma\}$$
$$= \sum C_S^\gamma \int_a^b dt_1 \int_a^b dt_2 \cdots \int_a^b dt_S\, f_S(t_1, \ldots, t_S) \quad (5)$$

where the C_S^γ denotes the number of various confluences of $(\gamma - s)$ infinitesimal intervals the maximum order of any confluence being $(\gamma - s)$. The C_S^γ coefficients being independent of the f functions can be derived from the following formula when the total number of entities is N,

$$N^\gamma = \sum C_S^\gamma N(N - 1)\cdots(N - S + 1) \quad (6)$$

a set of relations valid for $N = 1, 2, \ldots$

An alternative expression for these coefficients, obtained by Kuznetsov and Stratonovich,[6] is just the same as equation (6) in another closed form (as shown in Appendix I).

$$C_S^\gamma = \frac{1}{s!}\left[\frac{d^\gamma}{d\omega^\gamma}(e^\omega - 1)^s\right]_{\omega=0} = \frac{1}{s!}\sum_{k=0}^S \binom{S}{k} k^\gamma(-1)^{S-k} \quad (7)$$

Again for the connection between these product densities and Janossy densities the reader is referred to the handbook article by Ramakrishnan. An inverse relation between Janossy densities and product densities is detailed in Ref. 6.

Kuznetsov and Stratonovich have also introduced a very useful relation connecting the product density generating functional $L(u)$ defined below with the cumulants or the real correlation functions of the processes. To give an idea of these cluster functions and their relation to the product densities we write the following relations

$$f_1(t_1) = g_1(t_1)$$

$$f_2(t_1 t_2) = g_1(t_1)g_1(t_2) + g_2(t_1, t_2)$$

$$f_3(t_1, t_2, t_3) = g_1(t_1)g_1(t_2)g_1(t_3) + 3\{g_1(t_1)g_2(t_1, t_2)\}_{sym}$$

$$+ g_3(t_1, t_2, t_3) \tag{8}$$

The g functions are the cluster functions or the real correlations in any order. The product density generating functional defined as

$$L(u, T) = 1 + \sum_{n=1}^{\infty} \frac{1}{n!} \int \int f_n(t_1 \cdots t_n)u(t_1) \cdots u(t_n) \, dt_1 \cdots dt_n \tag{9}$$

can be expressed in terms of cluster functions g as follows:

$$L(u, T) = \exp\left[\sum_{n=1}^{\infty} \frac{1}{n!} \int \cdots \int_0^T g_n(t_1 \cdots t_n)u(t_1) \cdots u(t_n) \, dt_1 \cdots dt_n\right] \tag{10}$$

The product density functional leads directly to the probability generating function defined by

$$\sum_n P(n, T)z^n = h(Z, T) \tag{11}$$

The relation between $h(Z, T)$ and $L(Z, T)$ may be expressed as

$$L(Z - 1) = h(Z, T) \tag{12}$$

For more detailed discussion of these formulas the reader is referred to the article by Kuznetzov and Stratonovich in Ref. 6.

Employing product density techniques offers a very useful tool in the study of stochastic point processes and an excellent review of these procedures can be found in the article by S. K. Srinivasan.[7] As stated earlier we will now discuss the problem of photon statistics and correlations of ejected photoelectrons, and try to tackle it from the viewpoint of stochastic point processes.

3. COHERENCE IN OPTICS

Before taking up the actual problem of photon statistics, we will familiarize ourselves with the concepts of coherence in optics. The time honored concepts of coherence arise from the well-known Young's experiment. These concepts have recently been enlarged and enriched by the vast body of developments both in theory and optical technology. I am referring to the new inventions of masers and lasers. Although light waves were known to be of the same electromagnetic character as radio waves, the traditional types of optical sources were only generating noise. All the ingeneous techniques in optics were oriented toward a constructive use of noise. The invention of the optical maser facilitates the production of optical signals. Coupled with these advanced techniques in the field of generators of optical waves fast photodetectors have come into use which respond strongly to individual quanta of light.

To match this explosion in optical technology, the theory of optical coherence which till now dealt with phase correlations only at two spatial points was enlarged by the study of intensity correlations and correlations of higher order inaugurated by the Brown and Twiss experiments[1] which we shall study later. Since the phase and the number of a radiation field are conjugate variables, Fock representation which relates to the number in cells of momentum space is clumsy, when there exists coherences of higher orders in the field. Hence, Glauber and others[2] concentrated on the description of the fields in terms of coherent states, in which the number in each mode can vary. We shall not go into the field theoretic aspects of this problem, and to obtain some elementary notions we shall content ourselves by referring to the treatment of such states as minimum uncertainity wave packets, detailed in Schiff's quantum mechanics.

The concept of optical coherence has long been associated with interference because interference is the simplest phenomenon that recalls correlation.

Young's double-slit experiment will produce an interference pattern on the screen, provided there is a correlation in amplitude at two space–time points. This correlation can be called second-order coherence in view of the later developments like the Brown and Twiss experiment, which for the first time introduced intensity correlations. The criteria of coherence now got shifted from second-order to fourth-order correlations. Chaotic light or light from thermal sources are completely de-

scribed by second-order correlation while optical waves from other sophisticated sources can exhibit higher orders of correlations. To focus
our attention on the differences between different kinds of radiation
fields, let us review the necessary terminology of interference in optics.

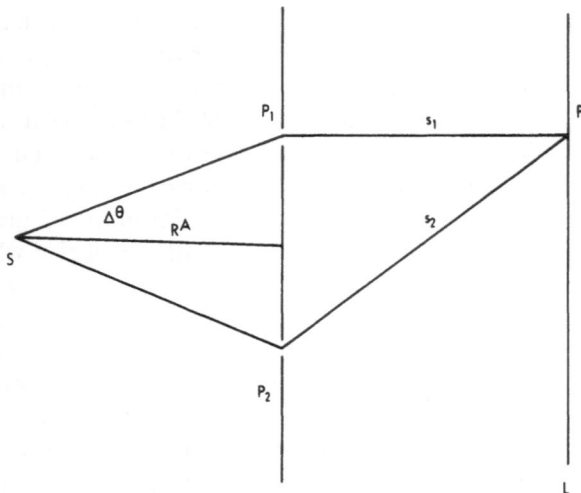

Fig. 1. Young's double-slit experiment.

Figure 1 depicts Young's double-slit experiment in which light
from a monochromatic source S gets split into two beams at the slits P_1
and P_2 and produces interference pattern on the screen L. The time delay
between the beams should not exceed $\Delta t = 1/\Delta\nu$ where $\Delta\nu$ is the natural bandwidth of the beam and this is called time of coherence and path
delay $c\Delta t$ corresponding to this is called length of coherence $l = c\Delta t$.
The area Δs of the source subtends at the pinholes an angle $\Delta\theta$ and for
producing the pattern they should be related as $\Delta s\Delta\theta \sim \lambda$ where λ is
the mean wavelength of the light used. This leads to the idea of area of
coherence $A \sim (R\,\Delta\theta)^2$. The volume of coherence is given by Al and the
average number of photons that are to be found in the volume of coherence is called the degeneracy factor δ, and this, for black-body radiation from a thermal sorce is given by Plank's formula $\delta = 1/(e^{h\nu/kT} - 1)$.
To appreciate the difference between thermal radiation field and modern
sources of optical waves like the laser or maser, we will consider and
compare quantities like $\Delta\nu, \Delta t, l$, etc., in the two cases. Now, $\Delta\nu$ for
thermal sources is about 10^8 cps, whereas for optical maser it is $< 10^2$
cps $\cdot l$ for thermal source is 1 meter, whereas it can be as big as 1000 miles

for maser light and, finally, the degeneracy parameter δ for thermal radiation is only 10^{-3}, whereas it is of the order of 10^{13} for coherent light beams like maser light. This enormous degeneracy factor makes it possible for us to describe these radiation fields in a semiclassical manner. To get an idea of the historical development leading to the development of these highly coherent beams we can summarize:

Einstein postulates stimulated emission	1917
Townes proposes maser as a source of *mm* wave	1951
Weber proposes microwave amplification	1953
Townes and Zeiger—Ammonia maser	1954
Bloombergen—Paramagnetic ion maser	1956
Townes and Schwlow—Optical region ruby maser	1958
Maiman—optical stimulated emission in solids	1960

In its broadest sense optical coherence theory deals with statistical fluctuation in the optical field. Therefore, the setting for the study of a stochastic point process is automatically introduced when we consider the amplitude or field variable $V(r, t)$ of the optical field (generated, e.g., by a thermal source) at a space time point r and t as a random variable, obeying the free Maxwell equations. As explained in the book by Born and Wolf (Chapter 10), it is advantageous to use instead of $V(r, t)$ the complex analytic signal $V(r, t) = \int_0^{\infty} V(r, v) \, e^{-2\pi i v t} \, dv$. Where $V(r, v)$ is the transform of the original real field variable. The observed intensity at a space time point (r, t) is $\langle V(r, t)V^*(r, t) \rangle$ where $\langle \cdots \rangle$ denotes the average over the proper ensemble. Quantum mechanical photodetectors measure this averaged product. Let us now define the correlation functions of amplitudes at two space–time points

$$\Gamma(r_1, r_2; t_1, t_2) = \langle V(r_1, t_1)V^*(r_2, t_2) \rangle \qquad (13)$$

This quantity by the assumption of stationarity and ergodicity will be given by

$$\Gamma(r_1, r_2; t_1, t_2) = \Gamma(r_1, r_2, t_1 - t_2) = \frac{1}{2T} \int_T^T V(r_1, t_1 + t)V^*(r_2, t_2 + t) \, dt$$

$$\text{as } T \to \infty \qquad (14)$$

For Gaussian light, i.e., chaotic light produced from thermal sources $\langle VV^* \rangle = \langle V^*V^* \rangle = 0$, since probability distribution for such cases if at all it can be given is assumed to be of the form $e^{-V(t+\tau/2)V^*(t-\tau/2)}$. Obviously, $\Gamma(r, r_1 0) = \langle |V^*(r, t)|^2 \rangle$, which is the intensity at the point (r, t).

The complex degree of coherence $\gamma(r_1 r_2 \tau)$ is defined as

$$\gamma(r_1, r_2, \tau) = \frac{\Gamma(r_1 r_2 \tau)}{\sqrt{\Gamma(r_1 r_1 0)\,\Gamma(r_2 r_2 0)}} \tag{15}$$

and therefore γ lies between 0 and 1.

With these definitions let us look at the way in which the interference pattern is produced in the Young's double-slit experiment. The total amplitude at P on the screen at time t, is the additive result of the amplitude at the two holes P_1 and P_2 at earlier times $t - S_1/c$ and $t - S_2/c$. The field at P on the screen at r and time t is given by

$$V(r, t) = \lambda_1 V(r_1 t - s_1/c) + \lambda_2 V(r_2 t - s_2/c) \tag{16}$$

where λ_1 and λ_2 are factors that arise from the geometry of the situation. Hence, we can calculate the intensity at P

$$\langle I(r, t)\rangle = |\lambda_1|^2 \langle I(r_1 t - s_1/c)\rangle + |\lambda_2|^2 \langle I(r_2 t - s_2/c)\rangle$$
$$+ 2\,\text{Re}\,\lambda_1^* \lambda_2 \langle V^*(r_1 t - s_1/c)V(r_2 t - s_2/c)\rangle \tag{17}$$

The last term in equation (16) is related to the modulus of the coherence function $\Gamma(r_1 r_2 (s_1 - s_2)/c)$ and varies as a cosine function of $(s_1 - s_2)/c$. Hence, calling the first two terms as I_1 and I_2, respectively, we can find the maximum and minimum values of I on the screen as $(s_1 - s_2)$ varies from point to point, as

$$I_{\max} = I_1 + I_2 + 2\,\Gamma_{12}$$
$$I_{\min} = I_1 + I_2 - 2\,\Gamma_{12} \tag{18}$$

We define the visibility of the fringes as

$$v = \frac{I_{\max} - I_{\min}}{I_{\max} + I_{\min}} \propto \left| \gamma\left(r_1 r_2 \frac{s_1 - s_2}{c}\right)\right| \tag{19}$$

where the coherence function is given by

$$\gamma = \frac{\Gamma_{12}}{\sqrt{I_1 I_2}}$$

when $\gamma = 1$ we obtain very sharp fringes. This criterion of coherence, defined by the visibility of the fringes only takes account of the second order correlation in amplitudes.

Michelson's Experiment

Another important use of coherence function or the visibility of fringes criteria is in the determination of angular diameters of stellar objects. Light from distant stars is reflected to movable mirrors and then by two fixed mirrors into the objective of a telescope. The interference fringes are observed on the focal plane. The coherence function can be shown to vary as a function of the distance $M_1 M_2$ (Fig. 2) between the mirrors. The vanishing of the fringes on the focal plane happens for a particular value of d given by $d = 1.22\,\lambda/\theta$, where θ is the angular separation of the star and λ the mean wavelength. For small values of θ, d has to be quite large and the optical system becomes unstable. The limit is already reached for $d = 6$m for $\theta = 0.2$ secs of an arc. Hence, interference fringes interferometry for these purposes has been abandoned in favor of intensity correlation interferometry

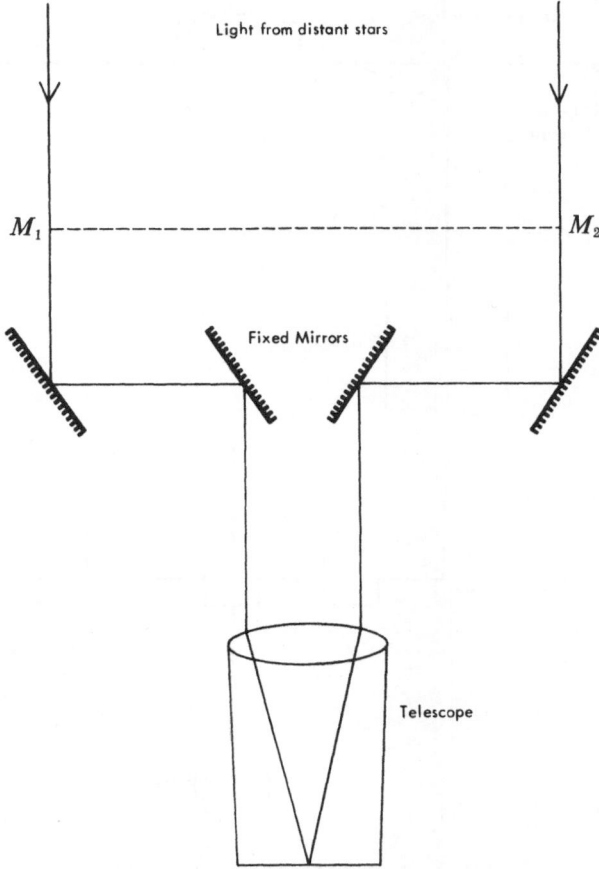

Fig. 2. Michelson's interferometer.

based on fourth order correlations, inaugurated by the Brown and Twiss experiments.

Brown and Twiss Experiment

In this experiment (Fig. 3) light from the source is collimated and split into two branches by a half silvered mirror each falling on a photodetector. The signals then are amplified, and get multiplied in a mixer or a correlator and then integrated in a recorder. Any amount of path or time delay can be introduced between the two branches. If the photomultipliers yield instantaneous currents $I_1 + \Delta I_1(t)$ and $I_2 +$

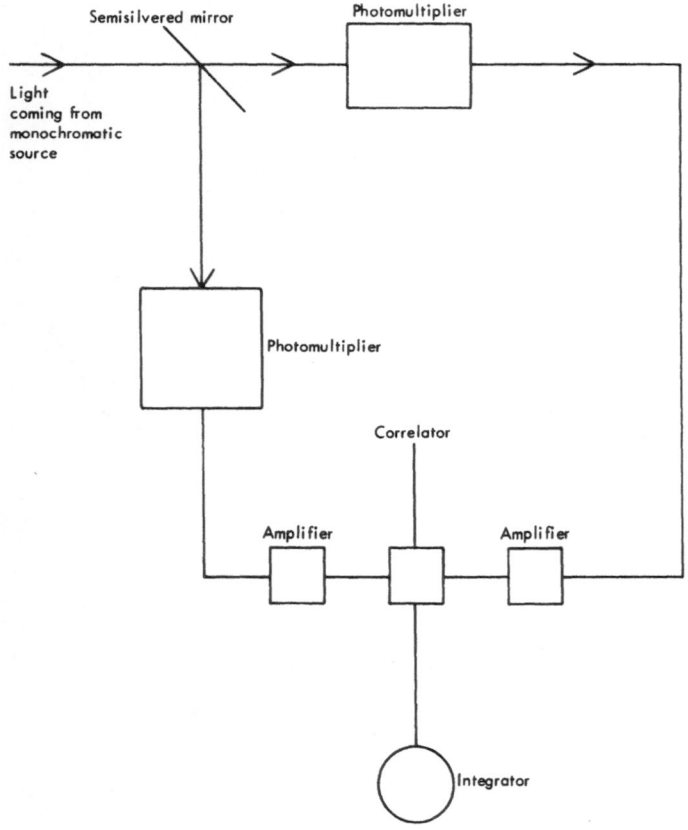

Fig. 3. Brown and Twiss experiment

$\Delta I_2(t)$, the average cross correlation function $G(d) = \overline{\Delta I_1 \Delta I_2}$ is studied. If the fluctuations are independent then this will be zero since ΔI_1 and ΔI_2 can separately take equally positive and negative values. It was seriously doubted whether there would be any connection between the coherence, in the wave and the correlations in the intensity of the two split beams. However, the result of Hambury, Brown, and Twiss experiment was positive. A correlation for the intensity fluctuation was exhibited commensurate with the wave coherence. A classical explanation based on photon statistics will be given in the next section. Since this set up does not require much phase stability, large stellar interferometers using this principle have been built in Australia, which operate on a base line up to 200 m, and angular diameters down to .0005 sec of an arc have been measured.

4. SEMICLASSICAL EXPLANATION

The above result can be analyzed from the general problem of fluctuation in statistical mechanics of a system which is in contact with a temperature bath. Taking the grand partition function $z = \sum_{N,i} e^{(\mu N - E_i, N_i)\beta}$ the average $\bar{N}^2$ is given by

$$\bar{N}^2 = \frac{1}{2\beta^2} \frac{\partial^2 z}{\partial \mu^2}$$

and

$$\overline{\Delta N^2} = \bar{N}^2 - (\bar{N})^2 = \frac{1}{\beta} \frac{\partial \bar{N}}{\partial \mu} \tag{20}$$

It is easy to see that for a classical Maxwell system $\overline{\Delta N^2} = \bar{N}$ and for the Bose–Einstein system

$$\overline{\Delta N_i^2} = \bar{n}_i(\bar{n}_i + 1) \tag{21}$$

This fluctuation is really more than in the case of the Maxwell gas. Suppose we have a Poisson distribution of particles and $P(n)$ the probability for having n particles, with the Poisson parameter λ we will have $P(n, \lambda) = e^{-\lambda n} \lambda^n / n!$ with the average value $\bar{n} = \sum nP(n, \lambda) = \lambda$. The fluctuation will be $\bar{n}^2 - \bar{n}^2 = \overline{\Delta n^2} = \lambda$. Thus we see that while the behavior of the classical Maxwell gas is similar to a Poisson situation, the fluctuations in a Bose gas are much more than Poisson arrival distribution. To see this more exactly we can also use the Einstein–

Fowler formula[8] for mean-square deviation $\overline{(\Delta E)^2}$ of the mean energy $\bar{E}$

$$\overline{\Delta E^2} = kT^2 \frac{\partial \bar{E}}{\partial T} \tag{22}$$

using Plank's law for the mean energy in an interval dv in a given volume in terms of the number of photons present therein.

$$E_v \, dv = \frac{8\pi h v^3}{c^3} \frac{v}{e^{hv/\kappa T} - 1} \, dv = \bar{n} h v \, dv$$

$$\overline{\Delta E^2} = \overline{\Delta n^2} \, h^2 v^2 \tag{23}$$

with $\overline{\Delta n^2} = \bar{n}(1 + \bar{n}\epsilon)$ where ϵ is the inverse of the number of cells in phase space in dv in volume V. In the experiment described above $\epsilon = \tau/t_0$ where ct_0 is the distance traveled by light during the observation time and $c\tau$ is the coherence length.

This excess fluctuation therefore arises from the bunching effect of the photons. This deviation from complete randomness makes it more likely that a photon assumes a certain set of phase–space coordinates if another photon is already there. Purcell's explanation of Brown and Twiss experiment will run as follows. Let the original n photons be split into n_1 into one branch and n_2 into the other. The fluctuation in the original beam

$$\overline{(\Delta n^2)} = \overline{(\Delta n_1 + \Delta n_2)} = \overline{(\Delta n_1)^2} + \overline{(\Delta n_2)^2}$$
$$+ \overline{2 \, \Delta n_1 \Delta n_2}$$
$$= \bar{n}(1 + \bar{n}\epsilon) \tag{24}$$

For each of photons the same fluctuation law holds. Therefore,

$$\overline{(\Delta n_1)^2} = \bar{n}_1(1 + \bar{n}_1\epsilon)$$

$$\overline{(\Delta n_2)^2} = \bar{n}_2(1 + \bar{n}_2\epsilon) \tag{25}$$

In view of equations (23), we arrive at the result

$$\overline{\Delta n_1 \Delta n_2} = \bar{n}_1 \bar{n}_2 t \tag{26}$$

which is never zero. Hence, the correlation is positive. For a Maxwell gas on this particle model, this will be zero. Correlation in the photon arrival counts is explained by the degeneracy feature or the clumping property the Bose particles. For thermal sources, long times of integration be necessary and the value of this correlation will also be reduced since the quantum efficiency of the photodetectors is less than unity. Moreover, if there is only partial coherence, equation (26) becomes

$$\overline{\Delta n_1 \Delta n_2} = \overline{n_1 n_2} \frac{\tau}{t_0} |\gamma_{12}(0)|^2 \qquad (27)$$

where $|\gamma_{12}|^2$ is the modulus square of the coherence function in the beam. In the next section we will translate these problems in the product density language and derive $P(n, \tau)$ the probability that n electrons are emitted in $(0, \tau)$.

5. STATISTICS OF PHOTON COUNTS

Let us consider the number of photoelectrons emitted from a photodetector in time 0 to T (Fig. 4). A knowledge of the statistical properties of the emitted electrons will give an insight into the nature of the light beam incident on the detector. If $I(t)$ is the intensity of the beam at time between t and $t + dt$ and if α is the sensitivity or quantum efficiency of the detector the average $\langle \alpha I(t) \rangle$ is the probability of

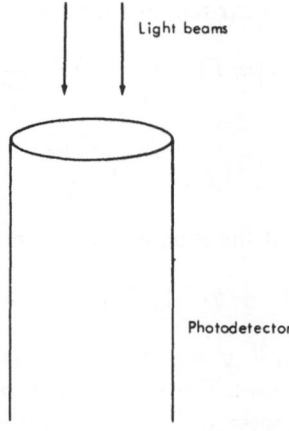

Light beams

Photodetector

Figure 4

finding one electron at time between $t, t + dt$. Alternately in the terminology of Section 2, it is the product density of finding a photoelectron between t and $t + dt$. If $\langle I(t) \rangle$ is a constant, i.e., I independent of time or if there is no correlation in the intensities at different times and I is a deterministic function of time. It is possible to write the probability $p(n, T)$ of obtaining n electrons in time 0 to T, as

$$P(n, T) = \frac{\left[\alpha \int I(t)\, dt\right]^n \exp\left[-\alpha \int_0^T I(t)\, dt\right]}{n!} \tag{28}$$

If I is allowed to be a random variable independent of t, but governed by a probability distribution $p(I)\, dI$, then $p(n, T)$ is given by

$$p(n, T) = \int p(I)\, dI \exp\left(-\alpha IT\right)\frac{(\alpha IT)^n}{n!} \tag{29}$$

Taking $p(I)$ corresponding to an uncorrelated Gaussian signal in the form

$$p(I) = \frac{I}{I_0} \exp\left(-\frac{I}{I_0}\right) \tag{30}$$

We obtain

$$p(n, T) = (1 + \bar{n})^{-1}(1 + \bar{n}^{-1})^{-n}$$

with

$$\bar{n} = \alpha I_0 T \tag{31}$$

This is a Bose–Einstein distribution with average number

$$\bar{n} = \sum np(n, T) = (1 + \bar{n})^{-1} x \frac{\partial}{\partial x} \sum_r x^n = \bar{n} \tag{32}$$

where

$$x = \frac{1}{1 + 1/\bar{n}}.$$

Similarly, the average of the square of the number is given by

$$\langle n^2 \rangle = (1 + \bar{n})^{-1}\left(x\frac{\partial}{\partial x}\right)^2 \sum x^n = 2\bar{n}^2 + \bar{n} \tag{33}$$

and the fluctuation is $\bar{n} + \bar{n}^2$ which is more than the Poisson by $\bar{n}^2$ This is characteristic of the Bose–Einstein distribution. If however, $I(t)$ is a correlated random process it is not possible to write an expression similar to (28). As explained in detail in Ref. 9, we will be dealing with a process with non-Markovian features. In the excellent review article of Mandel and Wolf[10] a bar over the entire equation (28) is drawn to denote the ensemble average and this will probably meet the needs of the situation. However, this difficulty can be overcome if we resort to the product density formulation of the problem.

    As pointed out earlier $\langle \alpha I(t) \rangle$ is the product density of observing an electron between t and $t + dt$ and, therefore, the mean number of electrons to be obtained in a given interval

$$\overline{n(0, T)} = \alpha \int_0^T \overline{I(t)} \, dt \qquad I(t) = V^*(t)V(t) \qquad (34)$$

and the mean square of the number in that interval as per that equation is

$$\overline{n^2(0, T)} = \alpha \int \overline{I(t)} \, dt + \alpha^2 \int_0^T \int_0^T \overline{I(t)I(t')} \, dt \, dt' \qquad (35)$$

Similarly, any moment of the distribution can be found using equation (5). If $f_2(t_1 t_2)$ can be written as

$$f_2(t_1 t_2) = f(t_1)f(t_2) + g(t_1 - t_2) \qquad (36)$$

then g relates to the coherence function in the beam. For

$$\overline{I(t_1)I(t_2)} = \bar{I}(t_1)\bar{I}(t_2) + |\Gamma(t_1 - t_2)|^2 \qquad (37)$$

if the amplitudes are distributed in a Gaussian manner. If in addition we assume that I's are independent of time and $|\Gamma(t_1 - t_2)| = \bar{I}$, we obtain the second moment of $p(n, T)$ as

$$\langle n^2 \rangle = \alpha \bar{I}T + 2(\alpha \bar{I}T)^2 = \bar{n} + 2\bar{n}^2 \qquad (38)$$

which is the second moment of Bose distribution.

From the second moment one can only determine the modulus of the coherence function Γ_{12}. We now propose that if one goes to the third moment one can find also the phase of the coherence function since the third moment can be expressed as

$$\langle n^3 \rangle = \alpha \int \bar{I}(t_1) \, dt_1 + 3\alpha^2 \int \overline{I(t_1)I(t_2)} \, dt_1 \, dt_2$$

$$+ \alpha^3 \int \overline{I(t_1)I(t_2)I(t_3)} \, dt_1 \, dt_2 \, dt_3 \qquad (39)$$

and

$$\overline{I(t_1)I(t_2)I(t_3)} = \overline{I(t_1)I(t_2)I(t_3)}$$
$$+ \bar{I}(t_1)|\Gamma(t_2 - t_3)|^2 + \bar{I}(t_2)|\Gamma(t_1 - t_3)|^2$$
$$+ \bar{I}(t_3)|\Gamma(t_1 - t_2)|^2$$
$$+ [\Gamma(t_1 - t_2)\Gamma(t_2 - t_3)\Gamma(t_3 - t_1) + c \cdot c] \qquad (40)$$

Evidently, the third moment provides a method of calculating the phase of the coherence also. For the Gaussian or themal light all correlations higher than the second can be expressible in terms of second-order correlations. Determination of moments up to third order gave both the magnitude and phase of the coherence function. Similarly, if a beam possesses coherence function up to lth order, we should go up to the $(l + 1)$th moment of the number obtained to determine all the coherence functions both in magnitude and phase. As is done Ref. 3, one can compare the moments with the moments of known probability distribution functions and get an idea of $p(n, T)$. However, if one has a knowledge of all the coherence functions of the light beam one can apply the methods of the stochastic process to actually get at the probability distribution functions using the equations (10) and (12) due to Kuznetzov and Stratnovich. They relate the product density generating functional to the g functions or the cluster functions which in the present context of optical coherence is the actual coherence functions which clearly stem from the deviations from the Poisson law. Hence, we can write $L(u)$ using the coherence functions $\Gamma(t_1, t_2, \ldots, t_m)$ in the place of the g function. If we are dealing with thermal light whose amplitudes are distributed in a Gaussian manner the mth order coherence is given by

$$\Gamma^m(t_1 \cdots t_m) = (m - 1)! \, \Gamma(t_1 - t_2)\Gamma(t_2 - t_3) \cdots \Gamma(t_{m-1} - t_m) \quad (41)$$

Mereover, if we make the assumption that $\Gamma(t_1 - t_2)$ is a constant given by the mean $(\alpha \bar{I} T)$, it is easy to show that product density generating function and hence, the generating function $h(z, T)$ of the probability $p(n, T)$ itself is given by

$$h(z, T) = [1 + \alpha \bar{I} T - \alpha \bar{I} z T]^{-1} \quad (42)$$

a result that identifies $p(n, T)$ as the Boson distribution. Our demonstration points to a powerful method that can be employed to arrive at $p(n, T)$ the probability distribution function if one is in possession of all the orders of coherence pertaining to the beam that is incident on the photodetector, giving rise to the photoelectrons.

6. PHOTOELECTRONS PRODUCED BY MIXING OPTICAL FIELDS OF DIFFERENT TYPES

Recently, many experiments are being performed with a view to determine $p(n, T)$ for the number of electrons ejected from the detector illuminated by a laser below and above the threshold. These bring out dramatically the transition of the laser action below and above the threshold. Near the threshold, a superposition of chaotic light with a coherent signal occurs, which can be considered as a rudimentary representation of a signal with a background noise. We are analyzing here a radiation field, in which the amplitudes of the Gaussian light is additively mixed with that of a Poisson field which is the signal. Many authors[11] have arrived at the resulting distribution using the P-representation of the superposed light beams. As illustrated in the earlier section the product density generating functional

$$L(u) = \left\langle e^{\alpha \int_0^T I(t)u(t)dt} \right\rangle$$

$$= \exp \sum \frac{1}{m!} \alpha^m \int \langle I(t_1) \cdots I(t_n) \rangle_c u(t_1) \cdots u(t_n) \, dt_1 \cdots dt_n$$

can be found by the cluster or cumulant functions. This expression can be generalized as was done by Kubo[12] to include several random variables I_A, I_B, ..., etc.

The product density generating functional $L(\{N\}, u, T)$ for N such random variables can be written as

$$L(\{N\}, u, T) = \exp K(\{N\}) \tag{43}$$

and

$$K(\{N\}) = \sum_{n=1}^{N} \sum_{\{n\}_N} K_n(\{n\}_N) \tag{44}$$

$K(\{N\})$ is the linked cluster sum over the exponential in the right-hand side expression of equation (43). $K_n(\{n\}_N)$ is the linked cluster sum corresponding to $\{n\}$ of $\{N\}$ total particles. If the mixture is one in which the amplitudes of chaotic light and Poisson light are additively mixed.

$$I(t) = [V_c^+(t) + V_s^+(t)][V_c(t) + V_s(t)] \tag{45}$$

where V_c refers to the chaotic field and V_s to that of the signal. Bearing in mind that the cumulants in (43) are non-zero only if all the variables are statistically connected and reduces to zero even if one of them becomes independent of others, one arrives at the generating functional:

$$L(u) = \frac{1}{1 - \lambda I_c} \exp \frac{I_s \lambda}{1 - \lambda I_c}$$

$$\lambda = \alpha \int_0^T u(t) \, dt \tag{46}$$

This is related to the generating function of the Laugerre polynomials. This method can be extended to any kind of mixing and the resultant generating function can be guessed if one knows all the correlations existing in the beam. For example if one mixes two separate Poisson signals with means $\bar{I}_{s_1}$ and $\bar{I}_{s_2}$ and a chaotic light with mean I_c, the resulting generating function will be

$$L(u) = \frac{1}{1 - \lambda I_c} \exp \frac{(I_{s_1} + I_{s_2})\lambda}{1 - \lambda I_c} \exp I_{s_1} I_{s_2} \lambda \qquad (47)$$

For more details the reader is referred to the preprint by S. K. Srinivasan and R. Vasudevan.[13]

In conclusion, our object has been to demonstrate that once a semiclassical representation of the coherent beams is possible, we can make use of the powerful methods of stochastic point process to realize many results that are obtained by other means, such as the making use of P-representations deduced from coherent state formalism governing these radiation fields. It may also be possible to devise interesting experiments bearing on the models of laser beams utilizing the techniques of stochastic theory illustrated in this paper.

APPENDIX A

Equation (6) can be re-expressed as

$$N^\gamma = \sum_{S=1}^{\gamma} C_S^\gamma \frac{N!}{(N - S)!} \qquad (A.1)$$

Substituting for

$$C_S^\gamma = \frac{1}{S!} \frac{d^\gamma}{d\omega^\gamma} (e^\omega - 1)^S \bigg|_{\omega=0} \qquad (A.2)$$

$$N^\gamma = \frac{d^\gamma}{d\omega^\gamma} \sum_{S=1}^{\gamma} \frac{(e^\omega - 1)^S}{S!} \frac{N!}{(N - S)!} \bigg|_{\omega=0} \qquad (A.3)$$

To (A. 2) we can add the additional terms with $S = 0$ and with S running from $(\gamma + 1)$ to N without changing the value of the expression since the differentiation $d^\gamma/d\omega^\gamma$, on the term and putting $\omega = 0$ later, will contribute nothing. In other words, we can write for the right-hand side

$$\frac{d^\gamma}{d\omega^\gamma} \sum_{S=0}^{N} \frac{(e^\omega - 1)^S N!}{S!(N - S)!} \bigg|_{\omega=0} = \frac{d^\gamma}{d\omega^\gamma} (e^\omega - 1 + 1)^N \bigg|_{\omega=0}$$

$$= N^\gamma e^{\omega N} \big|_{\omega=0} = N^\gamma \qquad (A.4)$$

Hence, $C_S^\gamma = d^\gamma/d\omega^\gamma (e^\omega - 1)^S \big|_{\omega=0}$. The expression on the right-hand side of (A. 2) by straightforward expansion equals

$$\frac{1}{S!} \sum_{k=0}^{S} k^\gamma \binom{S}{k} (-1)^{S-k} \qquad (A.5)$$

REFERENCES

1. R. Hanbury Brown and R. Q. Twiss, *Phil. Mag* **45:** 663 (1954) *Proc. Roy. Soc.* **242 A:** 300 (1957); **243 A:** 291 (1957).
2. R. J. Glauber, *Phys. Rev.* **130**, 2529; *Phys. Rev.* **131:** 2766 (1963). Sudarshan, E. C. G., *Phys. Rev. Letters* **10:** 277 (1963); *Proceedings of the Symposium on Optical Masers*, John Wiley & Sons, New York, 1963; Mandel L., Sudarshan, E. C. G., Wolf, E., *Proc. Phys. Soc. Lond.* **84:** 435 (1964).
3. L. Mandel, *Proc. Phys. Soc.* (London) **81:** 1104, (1963).
4. A. Ramakrishan, *Proc. Camb. Phil. Soc.* **46:** 595 (1950).
5. A. Ramakrishan, "Probability and stochastic processes" in *Handbuch der Physik*, **Vol. 4**, Springer Verlag, 1956.
6. P. I. Kuznetsov and R. L. Stratonovich, *Izvestiya Akad Nauk, Nauk, SSSR* Ser. Mat. **20:** 167 (1956). Translated in: *Nonlinear Transformations in Stochastic Processes*, Pergamon, London, 1965.
7. S. K. Srinivasan, *J. Math. Phys Sci*, **1** Nos. 1 and 2. p. 1, (1967).
8. A. Einstein, *Phys. Z* **10:** 185 (1909)
9. S. K. Srinivasan and R. Vasudevan, *Nuovo Cimento* **57:** 185 (1967).
10. L. Mandel and E. Wolf, *Rev. Mod. Phys.* **37:** 231 (1965).
11. R. J. Glauber, *Physics of Quantum Electronics*, McGraw-Hill Book Company 1966, 788.
12. R. Kubo, *Phys. Soc. Jap.* **17:** 1100 (1962).
13. S. K. Srinivasan and R. Vasudevan, MATSCIENCE preprint.

Stochastic Integration and Differential Equations— Physical Approach

S. K. SRINIVASAN

INDIAN INSTITUTE OF TECHNOLOGY*
Madras, India

---

1. INTRODUCTION

The theory of integrals of deterministic functions and solutions of differential equations, from the point of view of real analysis, is based on the fundamental notion of the limit of a sequence. With the advent of measure theory, Lebesgue integration, and generalized functions it has become possible to extend the idea of integration to the widest class of deterministic functions. It is therefore a reasonable question to ask whether random functions can be treated in a similar fashion. In fact, integrals of random functions which, for brevity, may be called stochastic integrals, arise from very many physical situations. Stochastic integrals are very well known to electrical engineers who very often have to deal with responses to random signals and noise. From a pure mathematician's point of view, limiting stochastic operations do not offer any special difficulty since all limiting stochastic operations follow from the notion of convergence almost everywhere, convergence in mean square and convergence in measure, provided we replace Lebesgue measure by probability and the ordinary space by the function space of random functions. While this analogy is useful in that it provides a sound mathema-

---

*Department of Mathematics.

tical basis for the formulation of probability problems, it does not enable us to compute quantities of physical significance. In fact, the situation is analogous to the theory of Riemannian integration where the evaluation of integrals is made by the use of primitives or by the use of Simpson's formula interpreting the integral as an area. We shall use an exactly similar method to interpret integrals of random functions. With this as objective it is convenient to introduce the notion of a realized trajectory of any random variable corresponding to a set of realized values of the random variable in any particular "experiment."

2. INTEGRALS OF RANDOM FUNCTIONS

If $x(t)$ is a random function of t, we introduce the realized trajectory of $x(t)$ which we denote by $x^R(t)$, where $y = x^R(t)$ is a curve we would obtain on a particular realization of the process if we plot x against t. Corresponding to this realization, we can obtain the integral $\int_a^b x^R(t)dt$. The fundamental question we would like to ask is, "what is the probability measure for $\int_a^b x^R(t)dt$ or for $x^R(t)$?" This is a difficult question and in fact, the problem is pathological in the general case. Again the notion of continuity is very subtle for random functions and we shall not go into it at the moment. We shall show presently that the problem can be tackled if we confine ourselves to a class of random processes $x(t)$ which are characterized by the following features:

1. $x(t)$ is a homogeneous Markovian process with respect to t (t being a one dimensional parameter with respect to which the process evolves or progresses); and

2. If $\pi(x|x_0; t)$ is the probability frequency function of $x(t)$ (the existence of such a function is a consequence of 1 so that $\pi(x|x_0;t)\,dx$ represents the probability that $x(t)$ takes a value between x and $x + dx$ given that at $t = 0$, $x(t) = x_0$, i.e., $\pi(x|x_0; 0) = \delta(x - x_0)$, $\pi(x|x_0;t)$ has the property;

$$\pi(x|x_0; \Delta) \simeq R(x|x_0)\Delta + \delta(x - x_0)\left[1 - \Delta \int_x R(x|x_0)\,dx\right] \qquad (1)$$

for all values of Δ. This is a fairly reasonable assumption. It is interesting to compare with the usual Gaussian approximation where the behavior of $\pi(x|x_0; \Delta)$ for small Δ is governed by

$$\int \pi(x|x_0; \Delta)(x - x_0)^n \, dx = O(\Delta) \qquad \text{for } n = 0, 1, 2$$

$$= o(\Delta) \qquad \text{for } n > 2 \qquad (2)$$

The above condition would not be satisfied for any process in which $\pi(x|x_0; \Delta)$ is itself expandable in powers of Δ as we have chosen above, since all the moments vanish only linearly in Δ. However we have been able to show if we choose a particular parametrization procedure for $R(x|x')$, $x(t)$ will describe processes like Brownian motion which on a first look does not seem to fit in with any mode of description.

   3. $\int_{x'} R(x'|x) \, dx' < \infty.$

   This is an important condition and, in fact, it is this that enables us to define a proper measure for the realized trajectory $x^R(t)$.

   Counter Example: Consider the passage of a high-energy electron in a medium like lead. Due to Bremsstrahlung and ionization loss, energy will change in a random manner. Let us assume that the ionization loss is a constant which is certainly the case for very-high energies. Radiation cross section as given by Bethe–Heitler is highly singular. $\int_{E'} R(E'|E) \, dE'$ diverges so that an infinite number of transitions in energy in a finite interval of t is possible. Our condition 3 prohibits it so that any finite interval there can be only a finite number of transitions. Both 2 and 3 together mean $x(t)$ remains a constant between any two transitions, the number of jumps being finite. A typical trajectory is given in Fig. 1. The probability measure for any general trajectory can be written easily. Defining

$$\alpha(x) = \int_{x'} R(x'|x) \, dx' \qquad (3)$$

as the total probability per unit t that a transition from x takes place, we note that the random variable which has assumed the value x_0 at t_0, jumps from x_0 to a value in $(x_1, x_1 + dx_1)$ between t_1 and $t_1 + dt_1$, to a value in $(x_2, x_2 + dx_2)$ between t_2 and $t_2 + dt_2$, $\cdots$ $(x_n, x_n + dx_n)$ between t_n and $t_n + dt_n$ is given by

$$\pi(x_n, x_{n-1}, \ldots, x_1|x_0; t_n, t_{n-1}, \ldots, t_1; t) \, dx_1 \, dx_2 \cdots dx_n \, dt_1 \, dt_2 \cdots dt_n$$

$$= e^{-\alpha(x_n)(t - t_n)} \prod_{i=0}^{n-1} e^{-\alpha(x_i)(t_{i+1} - t_i)} R(x_{i+1}|x_i) \, dx_{i+1} \, dt_{i+1} \qquad (4)$$

This is the probability measure for a typical trajectory given by

$$x^R(\tau) = x_0 + (x_1 - x_0)H(\tau - t_1) + \cdots + (x_n - x_{n-1})H(\tau - t_n) \quad (5)$$

where $H(\tau)$ is the Heaviside unit function. Next it is an easy matter to obtain $\pi(x|x_0; t)$;

$$\pi(x|x_0; t) = \sum_{n=2}^{\infty} \int_0^t dt_n \int_0^{t_n} dt_{n-1} \int_0^{t_{n-1}} \cdots \int_0^{t_2} dt_1 \int_{x_{n-1}} \cdots \int_{x_1}$$

$$\{ e^{-\alpha(x)(t-t_n)} e^{-\alpha(x_{n-1})(t_n - t_{n-1})} R(x|x_{n-1}) \prod_{i=0}^{n-2}$$

$$[e^{-\alpha(x_i)(t_{i+1} - t_i)} R(x_{i+1}|x_i) \, dx_{i+1}] \}$$

$$+ \int_0^t e^{-\alpha(x_0)t_1} e^{-\alpha(x)(t-t_1)} R(x|x_0) \, dt_1 + e^{-\alpha(x_0)t} \delta(x - x_0) \quad (6)$$

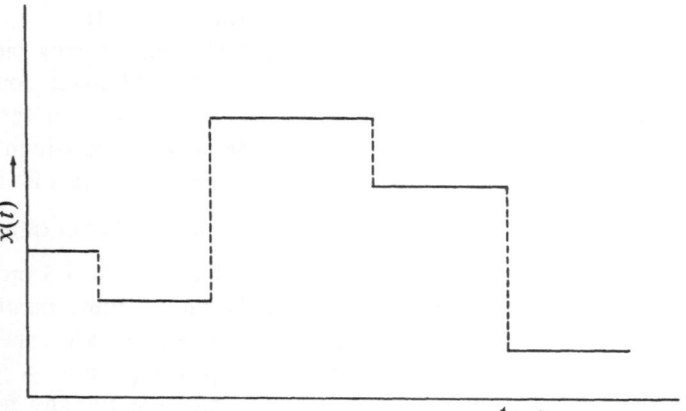

Fig. 1. Typical realized trajectory of a general basic random process.

We shall call $x(t)$ as a *Basic Random Process* (see for example References 1 and 2). The integral of $x(t)$ over any finite range (a, b) is easy to define. All that we have to do is to consider the class of realized trajectories $x^R(t)$ and the integrals $\int_a^b x^R(t)\, dt$ induced by them. The probability measure of

$$y^R(t) = \int_a^b x^R(t) \, dt \quad (7)$$

is the same as that of $x^R(t)$. This may appear too stringent a condition from a point of view of stochastic convergence. In fact it is not, and the interested reader can refer to the work of Ramakrishnan and

Vasudevan who have discussed this problem extensively. Thus in the present case, it is easy to define a stochastic integral and the corresponding probability measure.

Next question we would like to ask is, "Is this description adequate to explain, if not all, at least a good number of physical phenomena?" The answer is yes. We can build a number of nonconstant functions with finite number of jumps as follows: Consider $y_0(t)$ defined by

$$y_0(t) = \phi_0(t)x(t) \tag{8}$$

where $\phi_0(t)$ is a smooth function of t. The realized value $y_0^R(\tau)$ corresponding to a set of realized values $x_0, x_1, \ldots, x_n$ is given by

$$y_0^R(\tau) = x_0\phi_0(\tau) + (x_1 - x_0)\phi_0(\tau)H(\tau - \tau_1) + \cdots$$
$$+ (x_n - x_{n-1})\phi_0(\tau)H(\tau - t_n) \qquad 0 \leq \tau \leq t \tag{9}$$

The probability measure of y corresponding to $y_0^R(\tau)$ as given by (9) is given by (4). Once the probability measure of $y_0(t)$ is defined, the integral of $y_0(t)$ can be defined with a suitable probability measure. We can generalize (8) and define functions which have a higher degree of smoothness.

$$y_n(t) = \phi_n(t) \int_0^t y_{n-1}(\tau)\, d\tau$$
$$y_0(t) = \phi_0(t)x(t) \tag{10}$$

y_n is called a stochastic process of order n associated with the basic random process $x(t)$. The equation for $y_n^R(\tau)$, a typical realized trajectory of $y_n(\tau)$ is given by

$$y_n^R(\tau) = \phi_n(\tau) \int_0^\tau \phi_{n-1}(\tau_{n-1})d\tau_{n-1} \cdots \int_0^{\tau_1} \phi_0(\tau_0)x^R(\tau_0)d\tau_0 \tag{11}$$

The equations of realized trajectories are continuous for processes of order $n \geq 1$. The nth derivitive of $y_n^R(\tau)$ has finite discontinuities at a finite number of points in the interval $(0, t)$. The probability measure of a typical trajectory of $y_n(\tau)$ corresponding to an interval $(0, t)$ is the same as that of the corresponding trajectory of the basic random process. $y_n(t)$ is the most general random function we can construct. Of course, this does not exhaust all possible random functions. Equation (10) shows that the p.f.f. of $y_n(t)$ cannot be obtained without reference to $y_{n-1}(t)$ and the p.f.f. of $y_{n-1}(t)$ in turn cannot be obtained without any reference to $y_{n-2}(t)$ etc. However, this should not offer any special difficulty since we can deal with the joint p.f.f. of the set $y_i(t)$, $i = 0, 1, 2, \ldots, n$. We shall illustrate this presently by taking the Poisson's process. Alternatively we can write

$$y_n(t) = \int_0^t F_n(t, \tau) x(\tau) \, d\tau \qquad (12)$$

where $F_n(t, \tau)$ is given by

$$F_n(t, \tau) = \phi_1(\tau) \int_\tau^t \phi_m(\tau_m) \, d\tau_m \cdots \int_\tau^{\tau_4} \phi_3(\tau_3) \, d\tau_3 \int_\tau^{\tau_3} \phi_2(\tau_2) \, d\tau_2 \quad (13)$$

The realized value of $y_n(t)$ corresponding to the equation (5) is given by

$$y_n^R(t) = x_0 \int_0^t F_n(t, \tau) \, d\tau + (x_1 - x_0) \int_{t_1}^t F_n(t, \tau) \, d\tau + \cdots$$

$$+ (x_n - x_{n-1}) \int_{t_n}^t F_n(t, \tau) \, d\tau \qquad (14)$$

The p.f.f. of $y_n(t)$ can be obtained by first writing the joint p.f.f. of $y_n(t)$ and $x(t_i)$, $i = 1, 2, \ldots, n$ and then integrating over t_i.

The two methods described above are not always the convenient ones, since the integration and summation processes are tedious and do not lead to simple closed expressions for the p.f.f. of $y_n(t)$. We shall show how the p.f.f. can be arrived at in a simple and direct manner by dealing with specific examples.

3. SOME SIMPLE BASIC RANDOM PROCESSES

An example of a simple basic random process is provided by the Poisson process. The random variable assumes only the discrete values $0, 1, 2, \ldots, n, \ldots$ and the probability that $x(t)$ takes the value n is given by

$$\pi(n, t) = e^{-\lambda t} \frac{(\lambda t)^n}{n!} \qquad (15)$$

A typical realized trajectory of $x(t)$ is given in Fig. 2. Accordingly corresponding to the realization in Fig 2, (12) can be written as

$$y_m^R(t) = \int_0^t d\tau F_m(t, \tau) \int_0^\tau \frac{dx^R(\tau')}{d\tau'} \, d\tau'$$

$$= \int_0^t \frac{dx^R(\tau)}{d\tau} \phi_m(t, \tau) \, d\tau \qquad (16)$$

where

$$\phi_m(t, \tau) = \int_\tau^t F_m(t, \tau') \, d\tau'$$

It is interesting to investigate the nature of the realized trajectory of $dx^R(\tau)/d\tau$. This cannot be plotted graphically since the graph consists of the x-axis except at the points of discontinuities of $x^R(t)$ where $dx^R/d\tau$ has a delta function singularity. It is precisely this situation which makes the computation of p.f.f. of $y_m(t)$ fairly easy. If t_1, $t_2, \ldots, t_n$ are the points at which the Poisson events have occurred [i.e., $x(t)$ jumps by unity at every one of these points starting from $x(0) = 0$], then corresponding realized value of $y_m^R(t)$ is given by

$$y_m^R(t) = \sum_{i=1}^{n} \Phi_m(t, t_i) \tag{17}$$

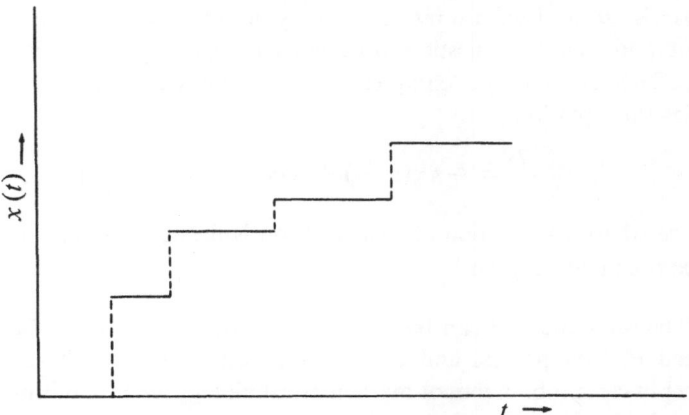

Fig. 2. A typical realized trajectory of a Poisson process.

If we denote by $\pi(y_m, t)$ the p.f.f. of y_m, then it is not easy to write the equation satisfied by $\pi(y_m, t)$ even though (17) looks harmless. Then there are two ways open.

(1) *Direct Method*:

The probability measure for $dx^R(t)/dt$ or $x(t)$ is given by

$$e^{-\lambda t_1} \lambda \, dt_1 e^{-\lambda(t_2 - t_1)} \lambda \, dt_2 \cdots e^{-\lambda(t_n - t_{n-1})} \lambda \, dt_n e^{-\lambda(t - t_n)}$$

Thus,

$$\pi(y_m, t) = \sum_{n=0}^{\infty} \lambda^n e^{-\lambda t} \delta \left[\int_0^t \Phi_m(t, t_n) \, dt_n \int_0^{t_{n-1}} \Phi_m(t, t_{n-1}) \, dt_n \right.$$
$$\left. \cdots \int_0^{t_2} \Phi_m(t, t_1) \, dt_1 - y_m \right] \tag{18}$$

The Laplace transform of $\pi(y_m, t)$ can be computed and it has an elegant closed expression,

(2) *Indirect Method*:

This is due to Ramakrishnan.[1] Define a variable

$$z_m(a, t) = \int_0^t \Phi_m(a, \tau) \frac{dx(\tau)}{d\tau} d\tau \tag{19}$$

where a is a parameter. Notice that $z_m(a, t) = y_m(t)$ when $a = t$. In view of this the p.f.f. of $y_m(t)$ and $z_m(a, t)$ are equal at the instant when $t = a$. However, the evolution of the two processes are different. In fact, $z_m(a, t)$ is a basic random process while $y_m(t)$ is a continuous random variable whose relationship to a basic random process is nontrivial. This method is useful if we are interested only in the p.f.f. at a particular instant and detailed questions like correlations of y_m at different t's cannot be answered. To complete the solution we note that $\pi(z_m, t)$ satisfies the equation

$$\frac{\partial \pi(z_m, t)}{\partial t} = -\lambda \pi(z_m, t) + \lambda \pi(z_m - \Phi_m(a, t), t) \tag{20}$$

Laplace transform solution of this equation is possible and the moments can be obtained very easily.

The same method can be applied to a variety of processes like generalized Poisson process and the Furry process. This may be of some interest in certain branches of mathematical biology where multiplicative process is of great importance.

Next we consider another basic random process in which $R(x'|x)$ the transition probability per unit time of a jump from x to x' is given by

$$R(x'|x) = R(x') \tag{21}$$

The transition probability is independent of its present value of the random variable. We note that $\pi(x|x_0; t)$ the p.f.f. of x at t given at $t = 0, x = x_o$ satisfies the equation

$$\frac{\partial \pi(x|x_0; t)}{\partial t} = -\alpha \pi(x|x_0; t) + R(x') \tag{22}$$

where

$$\alpha = \int_{x'} R(x') \, dx'$$

This is called a fluctuating density field and Ramakrishnan made extensive study of this process particularly in connection with some astrophysical problems. It can be shown that $x(t)$ describes a Gaussian random process provided (a) α is made very large, and (b) a suitable form of $R(x')$ is chosen. This has been discussed in detail in the Matscience Summer School lectures[5] with special reference to phenomena like Brownian motion, and I shall not go into it further at this time. The general integral corresponding to a realized trajectory can be written as

$$y_m^R(t) = \int_0^t \frac{dx(\tau)}{d\tau} \Phi_m(t, \tau)\, d\tau$$

$$= \sum_{i=1}^{n} (x_i - x_{i-1})\Phi_m(t, t_i) + x_0\Phi_m(t, 0) \tag{23}$$

We notice that the situation is exactly the same as in the previous case with one essential difference in the appearance of x_i's in the realized values since $x_i - x_{i-1} \neq 1$ in general. Now x_0 can be set equal to zero without loss of generality. Next we use the same technique employed before and define

$$z_m(a, t) = \sum_{i=1}^{n} (x_i - x_{i-1})\Phi_m(a, t_i) \tag{24}$$

We can readily write the differential equation satisfied by $\pi(z_m, t)$

$$\frac{\partial \pi(z_m, t)}{\partial t} = -\alpha\pi(z_m, t) + \int_x \pi(z_m - x\Phi_m(a, 0), t)R(x)\, dx \tag{25}$$

Equation (25) is an interesting equation in that it resembles a Poisson process even though $x(t)$ is much more complicated in its structure than a simple Poisson process. The impact of (24) on physical phenomena will be dealt with in the section on stochastic differential equations.

4. ALTERNATIVE APPROACH: POINT PROCESSES

Yet another different method can be used if we can fix our attention to what are called point processes, which arise whenever we have to deal with the distribution of discrete entities distributed over a continuous parameter. The parameter may stand for the vector velocity field of a turbulent fluid or the age parameter in a population. In such a case we introduce the cumulative number $N(x)$ of the entities taking parametric values $\leq x$, in which case $dN(x)$ is either 0 or 1 any other positive integer depending on the particular situation. The study of correlations of $dN(x_i)$ lead to what are called product densities and the recent work of Ramakrishnan and others have been discussed in the 1965 Matscience Anniversary Symposium.[6] The pertinent point which I want to project is that integrands of the type (23) or (17) can be viewed from the distribution of random points on the t-axis. With this motivation let us consider $y(t)$ defined by

$$y(t) = \int_0^t \Phi(t, \tau)\, dN(\tau) \tag{26}$$

with $dN(\tau)$ satisfying the properties mentioned above so that the integral can be written as a sum

$$y(t) = \sum_i \Phi(t, t_i) \tag{27}$$

Let us further assume that the point process is defined by the product densities $f_h(t, t_2, \ldots, t_h)$, $h = 1, 2, \ldots$ The moments of $y(t)$ can be easily evaluated by the application of the results obtained by Ramakrishnan[7] in connection with processes associated with random divisions of a line. If there are further random parameters in the integral as in the case of $y(t)$ defined by (24), even then explicit formulas are available for the moments of $y(t)$.

Apart from the evaluation of moments of $y(t)$, a more detailed picture of the process is provided by the correlation of $y(t)$ at different t's. To obtain these correlations, we note that if $y(t)$ is given by the stochastic integral (26), the second order correlation can be written as

$$\epsilon\{y(t_1)y(t_2)\} = \int_0^{t_1}\int_0^{t_2} \Phi(t_1, \tau_1)\Phi(t_2, \tau_2)\epsilon\{dN(\tau_1)\,dN(\tau_2)\} \tag{28}$$

Equation (28) follows from the commutability of expectation and integral operators. We next observe that

$$\epsilon\{dN(\tau_1)\,dN(\tau_2)\} = f_2(\tau_1, \tau_2)\,d\tau_1\,d\tau_2 \tag{29}$$

whenever $d\tau_1$ and $d\tau_2$ do not overlap and when they do,

$$\epsilon\{dN(\tau_1)\,dN(\tau_2)\}|_{\tau_1=\tau_2} = f(\tau_1)\,d\tau_1 \tag{30}$$

Using these results and taking into consideration the ranges of τ_1 and τ_2 we obtain

$$\epsilon\{y(t_1)y(t_2)\} = \int_0^{t_1}\int_0^{t_2} \Phi_1(t_1, \tau_1)\Phi(t_2, \tau_2)f_2(\tau_1, \tau_2)\,d\tau_2$$
$$+ \int_0^{\min(t_1,t_2)} \Phi(t_1, \tau)\Phi(t_2, \tau)f_1(\tau)\,d\tau \tag{30a}$$

Further interesting generalizations to higher-order correlations and the consequent break of particular coefficients that occur in the moment formulas, have been dealt with elsewhere in connection with the theory of Barkhausen noise.[8,9]

5. STOCHASTIC DIFFERENTIAL EQUATIONS

With the help of the techniques described in the previous sections, let us investigate the differential equation

$$a_0 \frac{d^n y}{dt^n} + a_1 \frac{d^{n-1} y}{dt^{n-1}} + \cdots + a_n y = x(t) \tag{31}$$

where $x(t)$ represents a basic random process the randomness in the equation arising from $x(t)$ only. There are two distinct cases according as whether the coefficients $a_0, a_1, \ldots, a_n$ are constants or functions of t. In the former case, it can be shown that $y(t)$ can be written as

$$y(t) = c(t) + \int_0^t F(t - \tau) x(\tau) d\tau \tag{32}$$

$c(t)$ can be set equal to zero without loss of generality in as much as we are concerned only with the statistical properties of $y(t)$. Then (32) reduces to an integral of the type given by equation (12) or equation (26) and all the results that have been derived in earlier sections are applicable to the present case. The relevant results relating to the correlation or equivalently the power spectrum which is more readily measurable in experiments have been derived for the general point processes by myself and Vasudevan. Since the results are new, they are given in Appendix.

Sometimes we may have to deal with systems of equation in which we have to reduce the simultaneous system to an equation of the type (31). In such a process of reduction, the derivatives of the function of the type $x(t)$ introduced, leading to inhomogeneous terms involving processes of different order. From purely physical considerations, it would be advantageous to combine functions representing processes of the same order since this is precisely the way in which they combine during the formation of differential equations. Thus a matrix formulation of the problem would provide a unified approach in the description of the statistical properties of the system as a whole. If we can generalize (31) to

$$\frac{d\mathbf{y}(t)}{dt} + A(t)\mathbf{y}(t) = \mathbf{x}(t) \tag{33}$$

where the components of $\mathbf{x}(t)$ can represent statistically independent basic random processes. The statistical independence of the different components of $x(t)$ can be removed in certain special cases.

The special case when the matrix $A(t)$ is independent of t can be treated in exactly the same manner as (31) with obvious changes in notation.[10] There is no general method of arriving at explicit solution of an equation of the type (32). However, we can recall a fundamental theorem which states that Φ is a fundamental matrix satisfying the homogeneous equation corresponding to (31) then the solution of y for zero initial conditions is given by

$$y(t) = \Phi(t) \int_0^t \Phi^{-1}(\tau)x(\tau)\, d\tau \tag{34}$$

Thus the solution for any particular y_i is of the form

$$y_i(t) = \sum_j \int_0^t F_{ij}(t, t')x_j(t')\, dt' \tag{35}$$

where

$$F_{ij}(t, t') = \sum c_{in}(t')c_j^{-1}(t') \tag{36}$$

(35) is useful in finding th correlation of y_i at different t's.

If we consider (33) directly, we can obtain the partial differential equation for the joint p.f.f. of $y_1, y_2, \ldots, y_n$, and $[x_i]$:

$$\frac{\partial \pi([y_i], [x_i], t)}{\partial t} = -\pi([y_i], [x_i], t) \sum_j \int R_j(x_j'|x_j)\, dx'$$

$$+ \sum_j \int_{x_j} \pi([y_i], [x_i]', t)R_j(x_j|x_j')\, dx_j'$$

$$+ \sum_k \left(-x_k + \sum_j a_{Rj}x_j\right)\frac{\partial \pi([y_i], [x_i], t)}{\partial y_k} \tag{37}$$

6. PHYSICAL EXAMPLES

Langevin's equation of the type

$$\frac{dy}{dt} + \beta y = F(t) \tag{38}$$

is an example of the first order random differential equation. (38) can describe the following:

1. The current $y(t)$ in a simple linear network, the applied electromotive force being the Thermal Noise Voltage $F(t)$ arising from the resistance,

2. The motion of a Brownian particle, the random force being $F(t)$, and

3. The fluctuation of voltage at the anode of a valve due to fluctuations in the number of electrons per unit time emitted by the cathode (Shot effect).

In case of (3), we can write equation (38) explicitly as

$$\frac{dV}{dt} + \frac{V}{Rc} = -\frac{\epsilon}{C}\frac{dN}{dt} \tag{39}$$

where N is a simple Poisson process. The problem is slightly more complicated if we have to be a little more realistic. This has been done and explicit evaluation of spectral density has been obtained.[8]

An example of a second-order differential equation is provided by an LCR circuit driven by noise voltage. The differential equation governing the process is given by

$$\ddot{y} + \beta\dot{y} + \omega_0^2 y = F(t) \tag{40}$$

where

$$\beta = \frac{R}{L} \qquad \omega_0^2 = \frac{1}{CL}$$

Third-order differential equations arise when we deal with (a) a suspended coil galvanometer subject to a random couple and random electromotive force, and (b) the Brownian motion of electrometers arising from thermal voltage fluctuation and random mechanical torque.

There are some physical phenomena which do not fit in exactly with the type of equations so far discussed. For the purpose of illustration we shall finally discuss briefly just two examples:

$$\frac{dc}{dt} = h(c_0 - c) - f(t)kc \tag{41}$$

where $f(t)$ is a random function of t. This equation arises in a model of emulsion polymerization. C is monomer concentration at any time and C_0 is the fixed monomer concentration in an ambient liquor (h: mass transfer coefficient governing the diffusion of monomer blob). The rate of polymerization is taken proportional to monomer concentration in the blob with the proportionality factor k. The stochastic element is $f(t)$ which has the values 0 and 1 corresponding to the starting and stopping of polymerisation at random times of arrivals from the ambient. More precisely,

$$f(t) = \frac{1}{2} - (-)^{N(t)}/2 \tag{42}$$

where $N(t)$ is the random variable governed by a Poisson process. The problem can be completely solved, if we define $\pi_0(c, t)$ and $\pi_1(c, t)$ as the joint p.f.f. and c and f. Then π_0 and π_1 satisfy the equations

$$\frac{\partial \pi_0(c, t)}{\partial t} = -(\lambda - h)\pi_0(c, t) + \lambda \pi_1(c, t) - h(c_0 - c)\frac{\partial \pi_0(c, t)}{\partial c} \quad (43)$$

$$\frac{\partial \pi_1(c, t)}{\partial t} = -(\lambda - h - k)\pi_1(c, t) + \lambda \pi_0(c, t)$$

$$-[h(c_0 - c) - hc]\frac{\partial \pi_1(c, t)}{\partial c} \quad (44)$$

with the initial conditions

$$\pi_0(c, 0) = \delta(c - c_0)$$
$$\pi_1(c, 0) = 0 \quad (45)$$

The relevant physical information can be obtained from the above equations, even though they cannot be solved explicitly. (For details see Ref. 11.)

(2) Scattering of a Schrödinger particle in a random potential.[12]

We shall assume that the potential is a function only of the radial distance. Now, ψ satisfies the equation

$$\psi''(l, k, r) + \left(k^2 - V(r) - \frac{l(l + 1)}{r^2}\right)\psi = 0 \quad (46)$$

where V is the random potential.

We also assume that the normal conditions imposed on the potential are satisfied almost certainly in the sense of stochastic convergence. Then we can arrive at the following differential equation for the phase shift:

$$\delta'(l, k, r) = -k^{-1}V(r)[\cos(\delta(l, k, r))J_l(kr)$$
$$- \sin \delta(l, k, r)N_l(kr)] \quad (47)$$

We can certainly obtain the Born approximate solutions in a formal way.

$$\delta(l, k, r) = -k^{-1}\int_0^r V(r)J_l^2(kr)\,dr \quad (48a)$$

or

$$\delta(l, k) = -k^{-1}\int_0^\infty V(r)J_l^2(kr)\,dr \quad (48b)$$

We can also write an improved Born approximate solution

$$\delta(l, k) = -k^{-1} \int_0^\infty dr V(r) J_l^2(kr) e^{\frac{1}{2k} \int_r^\infty V(r') J_l(kr') N_l(kr') dr'} \tag{49}$$

Finally, let us impose a simple restriction on $V(r)$:

$$V = \frac{\mu}{r}$$

μ is a random variable taking values ± 1, probability per unit distance of its translation from $+1$ to -1 and from -1 to $+1$ being p and q. We can set up differential equations for $\pi(1, r)$ and $\pi(-1, r)$

$$\frac{\partial \pi(1, r)}{\partial r} = -p\pi(1, r) + q[1 - \pi(1, r)]$$

$$= -(p + q)\pi(1, r) + q \tag{50}$$

with the condition

$$\pi(1, 0) = 1 \qquad \pi(-1, 0) = 0$$

$$\pi(1, r) + \pi(-1, r) = 1 \tag{51}$$

The mean value of μ is

$$\bar{\mu} = [(q - p)/(q + p)] + 2p \frac{2}{p + q} e^{-(p+q)r}$$

so that

$$\bar{V}(r) = \frac{q - p}{q + p} \frac{1}{r} + \frac{2p}{p + q} \frac{e^{-(p+q)r}}{r}$$

If we take a much simpler model, namely the one in which μ takes the value 1 or 0 with probabilities $\pi(1, r)$ and $\pi(0, r)$, we obtain

$$\overline{\mu(r)} = \frac{q}{p + q} + \frac{p}{p + q} e^{-(p+q)r}$$

Thus, the average value of V at any point r is given by

$$\overline{V(r)} = \frac{1}{r} \left(\frac{q}{q + p} + \frac{p}{p + q} e^{-(p+q)r} \right)$$

It is a pleasant surprise to note that the well-known Yukawa potential has a stochastic interpretation. Starting with a Coulomb potential μ/r we obtain the average potential which is partially Coulomb and partially Yukawa.

Still more striking is that if we put $p = q$ in the earlier model of attractive and repulsive forces alternately with equal probability, we get a type of screened Coulomb potential at every point.

We note that differential equation of the type

$$\sum a_i(t) \frac{d^i y}{dt^i} = x(t)$$

where a_i are also random functions has not been dealt with elegantly. Of coure, the techniques described here can be used to deal with problems of this type. The utility of the method in obtaining tangible results is to be known. Alternatively, one can write Fokker–Planck equation, this having the same status as our approach. Our experience with the two special types just now mentioned show that we may have to use indirect methods to extract results which can be directly correlated to experiments.

We have dealt with only a particular class of random differential equations that are amenable to explicit treatment by the techniques that have been developed here. For a more general treatment the reader is referred to the review articles of Professor Bharucha-Reid[13,14] and also a more detailed monograph by the same author.[15]

REFERENCES

1. A. Ramakrishnan, Phenomenological interpretation of the integrals of a class of random functions—I and II," *Proc. Kon. Ned. Akd. Von Wet.* **58** A: 471, 64 (1955).
2. A. Ramakrishnan, "A physical approach to stochastic processes," *Proc. Ind. Acad. Sci.* **44:** 428 (1956).
3. A. Ramakrishnan, "A stochastic field of a fluctuating density field," *Astrophys. J.* **119**:443, 682 (1954).
4. A. Ramakrishnan, "On stellar statistics," *Astrophys. J.* **122:** 24 (1955).
5. S. K. Srinivasan, "Fluctuating density fields," *Proc. Math. Sci. Sym.*, **5** (1966).
6. S. K. Srinivasan, "Sequent correlations in evolutionary stochastic point processes," *Proc. Math. Sci. Sym.*, **4**: 143 (1966).
7. A Ramakrishnan, "Stochastic processes associated with random divisions of a line," *Proc. Cam. Phil. Soc.* **49**: 473 (1953).
8. S. K. Srinivasan, "A novel approach to the theory of shot noise," *Nuove Gimento* **38**: 979 (1965).
9. S. K. Srinivasan and R. Vasudevan, "On a class of non-Markovian processes associated with correlated pulse trains and their applications to Barkhausen noise," *Nuovo Cimento* **41 B:** 101 (1966).
10. S. K. Srinivasan, "On a class of stochastic differential equations," *Zeit Ang. Math. Mech.* **43:** 259 (1963).
11. S. K. Srinivasan, "On the Katz Model for emulsion polymerisation. *J. Soc. Indust. Appl. Math.* **11:** 355. (1963).
12. A. Ramakrishnan, R. Vasudevan and S.K. Srinivasan, "Scattering phase shift in stochastic fields," *Zeit. Phys.* **196:** 112 (1966).

13. A.T. Bharucha-Reid, "On the theory of random equations. Stochastic Processes in Mathematical Physics and Engineering," A.M.S. (1964) 40-69.
14. A.T. Bharucha-Reid, "Semi-Group methods in mathematical physics," in: *Symposia on Theoretical Physics and Mathematics*, Vol. 2, Plenum Publishing Corp., New York, 1966, pp. 165-193.
15. A.T. Bharucha-Reid, *Random Equations*, Academic Press, New York, (1966).

[See the APPENDIX by S. K. Srinivasan and R. Vasudevan on the next page]

APPENDIX

*S. K. Srinivasan and R. Vasudevan**

Our object is to obtain the correlation of $y(t)$ as defined by (27) when y has attained stationarity for a general distribution of random points on the t-axis. We shall specialize (27) by assuming that ϕ is a deterministic function of $(\tau - t_i)$ and that the randomness occurs only in the form of a multiplicative parameter a_i whose distribution is assumed to be known so that

$$y(t) = \sum a_i F(t - t_i)H(t - t_i) \tag{A1}$$

where $H(t)$ is the Heaviside unit function. Thus, if the system attains stationarity, the second-order correlation will depend only on the difference $t_1 - t_2 = b$ and will be given by

$$\lim_{\substack{t_1 \to \infty, t_2 \to \infty \\ t_1 - t_2 = b}} \epsilon\{y(t_1)y(t_2)\} = \rho(b) = I_1 + I_2$$

$$I_1 = \lim \overline{a^2} \int_0^{\min(t_1,t_2)} f_1(\tau)F(t_1 - \tau)F(t_2 - \tau)d\tau$$

$$I_2 = \lim \overline{a^2} \int_0^{t_1} \int_0^{t_2} f_2(\tau_1, \tau_2)F(t_1 - \tau_1)F(t_2 - \tau_2)d\tau_1 d\tau_2 \tag{A2}$$

where we impose the restriction that F should tend to zero as its argument tends to infinity. The asymptotic form of $f_2(t_1 t_2)$ which we shall denote by $R(b)$ where $b = t_1 - t_2$ shall be assumed to be of such a nature that it tends to a constant as b tends to infinity. With such general qualifications for these functions we can obtain the Fourier transform of $\rho(b)$ in an elegant fashion as detailed below. The first order term in $\rho(b)$ is of the form

$$C = \int_0^{\min(t_1,t_2)} F(t - \tau)F(t_2 - \tau)f_1(\tau)d\tau \tag{A3}$$

If $t_1 > t_2$ and if $f_1(\tau)$ approaches a limit as $\tau \to \infty$, say $v/2$ the Fourier transform with respect to b becomes

---

*Faculty of Theoretical Physics, Matscience, Madras, India.

$$\tilde{C}(\omega) = 2 \text{ Re } \frac{1}{2\pi} \int_0^\infty e^{ib\omega} \, db \, \frac{\overline{a^2 v}}{2} \int_0^{t_2} F(t_1 - \tau)F(t_2 - \tau)d\tau$$

$$= 2 \text{ Re } \frac{\overline{a^2 v}}{4\pi} \int_0^\infty e^{ib\omega} \, db \int_0^\infty F(b + \tau')F(\tau')d\tau'$$

$$= \frac{v}{2} \, \overline{a^2} \cdot 2 \text{ Re } 2\pi |\tilde{F}(\omega)|^2 \tag{A4}$$

where

$$\tilde{F}(\omega) = \frac{1}{2\pi} \int_{-\infty}^\infty F(\tau)e^{i\omega\tau}d\tau \tag{A5}$$

The second-order term can be split into two parts A and B according as $\tau_1 > \tau_2$ or $\tau_1 < \tau_2$ and the integrals A and B and their Fourier transforms can be calculated as shown below

$$A = \bar{a}^2 \int_0^{t_2} d\tau_2 \int_0^{t_1} d\tau_1 f_2(\tau_1, \tau_2)F(t_1 - \tau_1)F(t_2 - \tau_2)$$

$$= \bar{a}^2 \int_0^{t_2} d\tau_2 \int_0^{t_1-\tau} d\tau \, R(\tau)F(t_1 - \tau_2 - \tau)F(t_2 - \tau_2) \tag{A6}$$

Changing the order of integration we get two terms

$$A = A_1 + A_2$$

$$= \bar{a}^2 \int_0^b d\tau \int_0^{t_2} d\tau_2 \, R(\tau)F(t_1 - \tau - \tau_2)F(t_2 - \tau_2)$$

$$+ \bar{a}^2 \int_0^{t_1} d\tau \int_0^{t_1\tau} d\tau_2 \, R(\tau)F(t_1 - \tau - \tau_2)F(t_2 - \tau_2) \tag{A7}$$

Taking the Fourier transform of the first term in the above with respect to b we get

$$A_T(\omega) = 2 \text{ Re } \bar{a}^2 \frac{1}{2\pi} \int_0^\infty e^{ib\omega} \, db \int_0^t d\tau \int_0^{t_2} d\tau_2$$

$$\times R(\tau)(b + t_2 - \tau - \tau_2)F(t_2 - \tau_2)$$

$$= 2 \text{ Re } \bar{a}^2 \frac{1}{2\pi} \int_0^\infty d\tau \int_\tau^\infty e^{ib\omega} \, db \int_0^{t_2} dx \, R(\tau)$$

$$\times F(b + x - \tau)F(x)$$

$$= 2 \text{ Re } \bar{a}^2 \int_0^\infty e^{i\omega(b'+\tau)} \, db' \int_0^\infty d\tau \int_0^\infty dx$$

$$\times F(b' + x)F(x)R(\tau)$$

$$= \bar{a}^2 \, 2 \text{ Re } 2\pi \, \tilde{R}(\omega)|\tilde{F}(\omega)|^2$$

where

$$\tilde{R}(\omega) = \int_0^\infty R(\tau)e^{i\omega\tau}\, d\tau \qquad (A8)$$

$\bar{a}$ and $\overline{a^2}$ are the average and mean square of the possible random quantities that can occur in the output function F due to each pulse. The second part of A is given by

$$A_2 = \bar{a}^2 \int_0^{t_1} d\tau \int_0^{t_1-\tau} d\tau_2\, R(\tau)F(t_1 - \tau_2 - \tau)F(t_2 - \tau_2)$$

$$= \bar{a}^2 \int_0^{t_2} d\tau^1 \int_0^{t_2-\tau'} d\tau_2\, R(\tau' + b)F(t_2 - \tau_2 - \tau')F(t_2 - \tau_2)$$

$$= \bar{a}^2 \int_0^\infty d\tau^1 \int_{\tau^1}^\infty dx\, R(\tau' + b)F(x - \tau')F(x) \qquad (A9)$$

Since

$$F(x - \tau') = 0 \qquad x < \tau' \qquad (A10)$$

we have

$$A_2 = \bar{a}^2 \int_0^\infty d\tau \int_0^\infty dx\, F(x - \tau)F(x)R(\tau + b) \qquad (A11)$$

The Fourier transform of this is given by

$$\tilde{A}_2(\omega) = \bar{a}^2\, 2\,\mathrm{Re}\, \frac{1}{2\pi} \int_0^\infty e^{ib\omega}\, db \int_0^\infty d\tau \int_0^\infty dx$$

$$\times\, R(\tau + b)F(x - \tau)F(x) \qquad (A12)$$

Concentrating on the integral B, corresponding to $\tau_1 < \tau_2$

$$B = \bar{a}^2 \int_0^{t_2} d\tau_1 \int_{\tau_1}^{t_2} d\tau_2\, R(\tau_2 - \tau_1)F(t_1 - \tau_1)F(t_2 - \tau_2)$$

$$= \bar{a}^2 \int_0^{t_2} d\tau_1 \int_0^{t_2-\tau_1} d\tau\, R(\tau)F(t_1 - \tau_1)F(t_2 - \tau_1 - \tau)$$

and as $t_1 \to \infty$ and $t_2 \to \infty$ this becomes

$$B = \bar{a}^2 \int_0^\infty d\tau \int_{b+\tau}^\infty dy\, R(\tau)F(y)F(y - b - \tau) \qquad (A13)$$

Remenbering $F(x) = 0$ when $x < 0$ the above integral becomes

$$B = \bar{a}^2 \int_0^\infty d\tau \int_0^\infty dy\, R(\tau)F(y)F(y - b - \tau) \qquad (A14)$$

Taking the Fourier transform of B we obtain

$$\tilde{B}(\omega) = 2\text{Re }\bar{a}^2 \frac{1}{2\pi} \int_0^\infty e^{ib\omega} \int_0^\infty d\tau \int_0^\infty dy$$

$$\times R(\tau)F(y - b - \tau)F(y) \tag{A15}$$

Combining this with Fourier transform of A_2 we get

$$\tilde{A}_2(\omega) + \tilde{B}(\omega)$$

$$= 2 \text{ Re } \bar{a}^2 \frac{1}{2\pi} \left[\int_0^\infty d\tau \int_0^\infty dx \int_0^\infty e^{ib\omega}\, db\, R(\tau + b)F(x)F(x - \tau) \right.$$

$$\left. + \int_0^\infty d\tau \int_0^\infty dy \int_0^\infty e^{ib\omega}\, db\, R(\tau)F(y - b - \tau)F(y) \right] \tag{A16}$$

$$= 2 \text{ Re } \bar{a}^2 \frac{1}{2\pi} \left[\int_0^\infty d\tau \int_0^\infty dx \left[\int_\tau^\infty R(b')e^{ib'\omega} \right] e^{-i\omega\tau} F(x)F(x - \tau) \right.$$

$$\left. + \int_0^\infty db'\, e^{ib'\omega} \left[\int_0^{b'} d\tau\, e^{-i\omega\tau} R(\tau) \right] \int_0^\infty dy F(y - b)F(y) \right. \tag{A17}$$

Remembering that real part of function F is also the real part of F^* we obtain interchanging the dummy variables, b' and τ in the second line of B,

$$\tilde{A}_2 + \tilde{B} = \bar{a}^2\, 2 \text{ Re } \frac{1}{2\pi} \int d\tau \int dx \int db'\, R(b')e^{-i\omega b'}\, F(x)F(x - \tau)$$

$$= \bar{a}^2\, 2 \text{ Re } (2\pi)\tilde{R}(\omega)|\tilde{F}(\omega)|^2 \tag{A18}$$

Thus, the power spectrum of the response can be written as

$$\tilde{A}_1 + \tilde{A}_2 + \tilde{B} + \tilde{C} = \tilde{F}(\omega) = 2\frac{\nu}{2}\,\bar{a}^2(2\pi)|\tilde{F}(\omega)|^2$$

$$+ 4\bar{a}^2(2\pi) \text{ Re } \tilde{R}(\omega)|\tilde{F}(\omega)|^2 \tag{A19}$$

Some Aspects of Operational Research

ALEC M. LEE

AIR CANADA*
Montreal, Canada

INTRODUCTION

In 1966, I published a book called "Applied Queueing Theory" which was well-received by my professional colleagues. One pleasant result was that Dr. Alladi Ramakrishnan, Director of the Institute for Mathematical Sciences in Madras, India, invited me to spend some time at the Institute as a visiting fellow to deliver a series of six lectures on operational research. The term of a visiting fellowship, three months, was too long to be acceptable to a non-academic person, such as myself, and for a time it seemed that the project would not be realized. However, Dr. Ramakrishnan devised a compromise by which I could present these papers in a three-week term as visiting scientist. This was planned for January 1968.

Operational research is not, of course, new to India. Various government agencies and the Armed Forces have been active in applying it for some years, and the Operational Research Society of India is growing steadily both in membership and influence. However, in south India (of which Madras is the principal city) operational research is a fairly recent development and, except for a few industrial companies—particularly in the textile field—it so far has been little applied. Consequently, Dr. Ramakrishnan requested that my papers be divided between general discussions of the subject (which could interest the

---

*Director of operational research.

industrialists and businessmen of the city) and some of its technical applications (which would interest members of the flourishing academic community). Broadly speaking then, sections 1, 2, and 6 are general in nature while Sections 2, 4, and 5 consist of more technical material chosen to illustract the general points made in the others.

1. AN OVERVIEW

Operational research is often said to be an applied science. Perhaps it would be more accurate to say it is an application of the methods of science and the techniques of mathematics to problems arising in one particular field of human activity. This field is that of organized, purposive human endeavor: industry, commerce, government, and warfare. In other words, operational research concerns the behavior of man-made, organized systems of men, machines, materials, and money which possess—in the words of Ackoff and Rivett—content, structure, communication, and control. The subject matter of operational research is then decidedly extensive and important. We shall see in the sequel to what extent its methods may be called scientific.

The people who engage in O. R. are not, for the most part, interested merely in analyzing and understanding the way in which organized systems behave. In this way they differ from, say, astronomers who are interested in understanding and predicting the behavior of certain natural, physical systems. They are often, perhaps even mostly, more like engineers in applying the results of scientific work to the development or improvement of those organized systems of concern to them.

However, it would be improper to suggest that the practice of O. R. does not generate the need for new theory and new techniques. It does. Some practitioners and many academic persons do apply themselves usefully to satisfying these needs. Whether or not the work which they do in this regard may be classified as *theoretical* O. R.—itself an apparent contradiction in terms—it is difficult to say. Perhaps most of it may be assigned to other existing fields of research: probability theory, say, or combinatorial analysis. In these seminars we shall take this view. *If* there is genuinely a theoretical discipline of which O. R. is the practice it is probably Cybernetics, the "science of communication and control in man and the machine," as Wiener defined it. For my own part, however, I find this too limiting for, as we shall see, much of what most practitioners would call O. R. is concerned with other matters than the communication and control aspects of organized systems.

Organizations and Systems

I do not want to claim too much for my profession. Men have been actively studying and designing organized systems for a long time indeed—certainly long before the inception of O. R., thirty or so years ago. In some parts of the world men began to gather and found settled communities (a form of organized system) about ten thousand years ago. Jericho, as the archeological work of Professor Kathleen Kenyon has shown, goes back in its earliest incarnation almost that far. There are no written records, of course, but the archeological evidence suggests that to have accomplished so much Jericho must have been an organized system like many others. We may postulate that it could be found to comprise three types of organized subsystems.

The first of these we shall call *operational systems* (dropping the prefix sub- where no confusion can result). An operational system is a coordinated assembly of men, tools (or machines) and power-sources such as horses, oxen, or engines which do the physical work of converting inputs of materials and energy into outputs of goods, artifacts or services. I stress the term *physical work*, for these are systems for doing things—for, if you like, "bringing home the goods." A farmer with his oxen and a wooden plough make up an operational system. It is admittedly a small system, but, in point of fact, it is not a simple one. In it there is a controlling element—the man—at the level of a *steersman*. Similarly a group of four men running a highly-instrumented, modern steel furnace is an operational system, though again a fairly small one.

The second we shall call an *information system*. In some ways an organization's information systems correspond to a man's nervous system. We shall not pursue the intriguing possibilities of this analogy further in these seminars; they are being adequately traced by others. In the round, so to speak, an information system consists of a coordinated assembly of men, machines or tools, and materials which collects inputs of data, stores them, processes them into information outputs or reduced data, and distributes these outputs to other subsystems of the whole organization. It is difficult to imagine any organized, man-made system existing without information subsystems. For example, records are extant from early Mesopotamian temples of 5000 years ago showing the amounts of grain, sheep and other income in kind delivered to, or distributed from, the temple warehouses. This was an early example of a type of information system that is common today—an *inventory control system*—and much studied by O. R. practitioners. Governments from very early times have set up systems to inform them of the number of their subjects, their ages and their wealth to permit judicious decisions to be taken in the matter of taxation. The Babylonians had systems to provide information of this sort. We have them today not only to take regular censuses of population, but also of production, investment, imports, exports, etc.

Finally, the third type of subsystem we may call *management systems*. This is not perhaps a well-chosen term, but it is not easy to find a better. The word management refers to the main control mechanism of the organized system. Physically, a management system must consist of men, equipment, and *concepts*, acting together (hopefully) to shape the future course of the organization to which it belongs: framing policies, stating objectives, and specifying means for accomplishing them. Management systems are linked formally to operational systems reciprocally by information systems. Informal links commonly exist as well. The structure of management systems, and the processes by which they perform their functions, have been subjects of inquiry and speculation for millenia. Systematic studies and design proposals of considerable antiquity—such as are sketchily outlined in Plato's *Republic*—are still extant. In recent years, management systems and the elite groups who tend to man them have been investigated intensively by sociologists.

These remarks have been, enough to show that man-made organized systems—or *organizations* for short—and the subsystems of which they are composed have been the subject of serious consideration for a long time. However, I suggest that the manner in which such systems were studied was, until quite recent times, certainly unscientific and even perhaps unsystematic. Their design suffered as a consequence. What we recognize nowadays as the *scientific method* of making investigations of the properties and behavior of systems had its true origin not much more than 350 years ago in the writings of Sir Francis Bacon. Since 1620 (the data of Bacon's *Novum Organum*) the scientific method has been developed and extensively applied to the understanding of natural systems, e.g., the solar system, biological systems, and physical systems such as enclosed gases. Its success has been remarkable.

The Scientific Method

What is the nature of this scientific method?   In the first place it is a generalized procedure for studying the properties and behavior of natural systems. It does not say specifically what a scientist must do at any point in time to solve a problem. It is not a series of programable rules which if fed to a computer or an automaton would permit it to conduct scientific inquiries. In the second place, it is an unending iterative procedure based upon the premise that no result or theory of science can ever be final. No statement about any natural system can ever be shown to be absolutely true; it can at best only not have proved false. It is summed up in terms of five stages or activity, as follows:

1. *System Definition*: define the boundaries and content of the natural system or process to be studied. (In most cases redefinition at later stages is needed.)

2. *Data Collection*: by means of observation of the dynamically evolving system (e.g., in astronomy) or by direct, controlled experimentation (e.g., in physics) assemble facts and data. Form tentative classfications, make tentative analyses: collect more data as then indicated.

3. *Form Hypotheses*: analyze data further to indicate the form and nature of imbedded relationships. Form a hypothesis or theory to explain observed phenomena or suggested relationships.

4. *Devise Tests*: consider what phenomena that are observable and measurable the theory can predict, which, if the predictions are not supported, are significant enough to reject the theory. (Tests should be framed so as to stress rejection in the case of failure, not acceptance in the case of success.) Design experiments of observations.

5. *Execute Tests*: make the test experiments or observations, analyze the results and accept or reject the theory. Then: a. If rejected; return to stage 1 to redefine the system, or to stage 3 to frame new hypotheses. b. If accepted—temporarily, for all results in science are *interim* results—expand the system boundary, or focus attention on other content of it, and re-enter stage 1.

While every one of these stages is significant and has raised problems of philosophy or technique or both, the third has a particular importance. This is the stage which specially expresses the scientist's *Weltanschauung*—his mental picture or reality—and consequently influences his choice of models of the systems and processes which he is concerned with. For example, the period of scientific development which began about 1650 and lasted until about 1860 coincided with the first great period of advances in mechanics. Pendulum-regulated clocks, helical-spring driven chronometers regulated by balance-wheels, steam-engines and so on, all data from the early part of that period. A mechanistic view of nature was dominant and scientific theories were on the whole predicated upon mechanistic models. Newton's is a "clockwork" universe. It is also a deterministic one, expressible in symbolic, mathematical terms. Developments in the biological sciences from the mid-19th century onwards brought about statistical and probabilistic views of nature which were reinforced by developments in physics such as the kinetic theory of gases and Planck's quantum theory. The world as a stochastic process was a view soon to be accepted.

In the application of the scientific method to the analysis, understanding, improvement and design of man-made, organized systems—*organizations*, as we shall call them—there have been similar shifts in viewpoint. It is of interest to note what some of these have been so that the antecedents of operational research may be recognized. We cannot trace all of them here. The family tree is not always clear, but we can point to some of them.

The Emergence of "Scientific Management"

The 250 years following 1620 constituted the first great period of application of the scientific method in science. It was not a period of great application in industrial, commercial or governmental organizations. An exception was the work of Perronet, who in 1760 carried through some systematic analyses of the manufacture of pins. This, reported by Diderot in the *Encyclopédie*, gave Adam Smith the inspiration for a celebrated passage in his *Wealth of Nations* in which he discusses the division of labor into 18 distinct operations in making a single pin. The same example was quoted by Babbage in his book of 1832 *On The Economy of Manufactures* in which mathematical approaches to the analysis of factory operations are dealt with. It is noteworthy that Babbage was also the first person to devise and attempt to construct what we should recognize as a computer.

The application of rational methods to the operations of organizations really began in earnest in 1881. In that year F. W. Taylor began his series of scientific studies of production processes in steelworks. In 1903 he began to publish papers and books which outlined his ideas on the application of the principles of scientific investigation to the solution of business problems and so inaugurated the period of so-called scientific management. This term is misleading. The approach of Taylor and his contemporaries Frank and Lillian Gilbreth and their followers was systematic rather than strictly scientific; and the problems and processes which they addressed themselves to were usually those at the shop-floor level rather than those of management. Taylor, however, clearly realized and stated that he was using analytic processes very similar to those of the scientific method.

It is important to recognize that an organization was implicitly viewed by Taylor and the devotés of scientific management as a *machine*. In adopting this viewpoint they were merely following in the wake of the majority of 18th and 19th century scientists whose models of natural systems were, as we have noted, almost all at the level of clockworks. Newton's is a clockwork universe; Carnot's universe is a simple heat-engine; Adam Smith's is a clockwork economy; and Taylor's industrial organizations are clockworks right down to the individual persons in them. This is, to be sure, a somewhat limited view of reality and we cannot be surprised that its forceful application by means of time-and-motion, studies and rate-fixing (based with unusual inappropriateness on the medieval concept of the *just price*) should have been received with uncertain enthusiasm.

At the same time, as a result of the work of a number of consultants interested in the management structure—or organization—of organizations since the first World War, it was being accepted by the more adventurous managers that a rational approach to such matters might not be harmful. The work of the prominent theorists of organization structures—of whom Urwick and Fayol are outstanding names—was not scientific in the sense that it was much based upon the strict application of the scientific method; but it undoubtedly was founded on rational speculation, broad reading and much practical observation. It helped create a receptive climate. It was also undeniably useful; though due (as Urwick in particular was quick to point out) to numerous mis-understandings of the competent work of theorists of military organization (particularly in Germany), its success was sometimes in doubt. The predominant view of an organization in all of this work was that of an hierarchical command structure, with formalized internal channels of information flow. Urwick again, drawing upon military experience, was foremost in pointing to the need for, and inevitability of, informal communication in an efficient organization.

In the late 1920's, as a consequence of some work undertaken by the Harvard psychologist Elton Mayo in the plants of the Western Electric Company at Hawthorne, near Chicago, a different view of industrial organizations was added. This is the view of an industrial organization as a *social system*. Mayo pointed out that human beings are human, natural systems of an indeed somewhat high level of complexity. This startling revelation was at first greeted by experienced managers generally with incredulity; but when Mayo showed that if they would act as if it were true (whether they believed it or not) their profits would rise, their doubts were replaced by enthusiasm. The importance of Mayo's work cannot be underestimated.

Another important antecedent development was the introduction in the mid-1920's by another American, W. A. Shewhart, of the technique of Statistical Quality Control (SQC). This derived from earlier work in mathematical, statistics itself resulting from the analytical needs of the biological sciences. The names of Galton, Edgeworth, Karl Pearson, W. S. Gosset and R. A. Fisher are particularly associated with it. One point must be made, however: SQC was an early, scientifically-based formalization of the principle of *management-by-exception* to ongoing processes. That is, management states targets or standards for some operational subsystem and arranges for it to be monitored: only if performance deviates from these by more than assigned limits is the matter reported and action taken. This is an important means of reducing the flow of redundant information in an organization.

Related to SQC, and now much more important, was the development in the late 1920's by E. S. Pearson and J. Neyman of the Theory of Statistical Testing of Hypotheses. The antecedents of this theory are numerous: it goes back to Thomas Bayes' posthumous paper of 1763 in which Bayes' Theorem was first stated. This work has been extended in recent years into Statistical Decision Theory, an important tool in the design of management systems. During the 1930's, in Britain particularly, these matters stimulated controversy and controversy led to fairly widespread interest.

The Birth of Operational Research

Such then was the state of affairs in the mid-1930's. The notion that systematic, and if possible scientific, approaches to the solution of the problems of industrial and other organizations could be useful was taking root—particularly, perhaps, in the United States. Surprisingly, however, the next step—and from our viewpoint here the important one—was taken in Britain. In January 1935, inconspicuously, a new committee met in London under the auspices of the British Air Ministry; "To consider how far recent advances in scientific and technical knowledge can be used to strengthen the present methods of defence against enemy aircraft." This was to be the origin of operational research.

I do not propose to describe the work and make-up of the Tizard Committee as it became called; this has been done with authority and skill by P. M. S. Blackett who was on it, and by C. P. Snow who followed its affairs with interest. Out of it came, in 1937, the first formal O. R. group at Biggin Hill; out of the committee's endeavors and the new O. R. groups came the integrated "weapons system" of Spitfires, Hurricanes, air-to-ground radio-telephones, radar stations, observer posts, operations control centers and the central control of Fighter Command, the trained personnel and the understanding that won the Battle of Britain in 1940. Instead let us conclude this lecture by identifying some of the principal characteristics of operational research as it developed in wartime—first in Britain, later in Canada and the United States—so that in the sequel we may trace its further evolution against a bare framework of reference.

1. O. R. makes use of the scientific method in analyzing problem situations and developing solutions to them.

2. O. R. is carried out by teams of people with varied backgrounds, (such as engineers, mathematicians, physicists, economists and psychologists) who will look at any organized subsystem, or process or an organization itself from different viewpoints. It is unlikely that all will see it as a clockwork, for example, or as a social system. Overall the team will see it as a total system. Any specific problem will be resolved by means of the most appropriate technique, or combination of techniques, irrespective of the disciplines where they have originated.

3. O. R. does not concentrate only upon problems in the operational systems of an organization, but is just as applicable to those in its information systems. (At present not much has been done on those matters that are peculiarly part of its management systems, i.e., the evaluation of policy, the development of long-range plans, etc., but a start is being made.) It is equally applicable to the *interface* problems between operational and information subsystems. It is therefore a general field.

4. O. R. functions within an organization as an *advisory* service to management: for successful O. R. to be done it is essential that communication between the O. R. personnel and the various management groups that may exist to be easy and open.

It is now appropriate to consider some applications of operational research and I shall, in the next section, describe some problems in the design of some operational subsystems.

2. SOME OPERATIONAL PROBLEMS

When industrial O. R. began after the second World War had ended, it found relatively quick acceptance in Britain in two big areas. First, due to the influence of the new director, Sir Charles Goodeve, of the British Iron and Steel Research Association the steel industry began to make use of O. R. methods. At the beginning of the 1950's, individual British steel companies such as the Steel Company of Wales began to establish their own O. R. departments. Second, due to the number of experienced O. R. scientists leaving the Royal Air Force and, to a lesser degree, the Royal Navy who joined aviation organizations, the two major airlines, BOAC and BEA, as well as the Ministry of Civil Aviation became early interested in operational research. Now it is a characteristic of both of these industries that many of the most apparent operational problems which they face concern queueing and congestion processes. Furthermore, in the early years of industrial O. R. one of the few, obviously useful bodies of mathematical theory already in existence was the so-called *Theory of Queues*. This existed as the result of investigations carried out by telephone engineers in several countries but notably Denmark (Erlang), Britain (Crommelin), the United States (Molina) and Germany/France (Pollaczek). The formulas available as a result of this earlier work were, moreover, not difficult to apply; the computation required was moderate. In a period such as 1945–55 when electronic digital computers were rare indeed this was a considerable advantage. It is not surprising therefore that queueing problems for a time became undoubtedly a major interest of O. R. workers.

It is no longer, I think, the most important preoccupation of those of us working in the field of operational research. Since 1955, O. R. has been gradually but steadily accepted in an increasing range of industries and government departments. In many of them there are problems of greater significance than queueing: the evaluation of capital investment projects, say, or the planning of advertising campaigns. Moreover, the great increase in the size and numbers of computers available to O. R. workers has made it possible for many types of hitherto intractable problems to be resolved. Last but not least, mathematical models and computational algorithms applicable to a greater range of practical problems have been developed and publicized. Nevertheless, the solution of practical problems of congestion and queueing is still an active occupation of O. R. workers in industry and some public sectors. The Port of New York Authority still functions and operates bridges, tunnels, and airports as it did 12 ago when L. C. Edie directed his classic studies of queueing at toll-booths. The British Airports Authority and the Canadian Department of Transport must still plan and operate airports where queueing problems inevitably arise as in the early days of Pearcey and Bell.

I propose to illustrate this field by describing some aspects of three queueing problems which I worked upon in collaboration with P. A. Longton just ten years ago in England. These problems arose in connection with the planning of a new city passenger airline terminal in London. They were: the problem of the size of the taxicab set-down curb, the problem of the number of check-in counters (both relating to departure operations) and the problem of the short-term car park (both arrival and departure operations). There was an existing terminal on the same site designated for the new one, and this was studied to obtain much of the necessary data. Let us begin by defining the whole system with which we were concerned: a city airline passenger terminal.

In many European cities an airline passenger who wishes to join a flight for which he holds a reservation has two choices. He can either make his own way to the airport and check in for his flight there, or he can go to a city airline terminal, check in there and be conveyed to the airport by the airline as a fully accepted passenger. The city terminal is not merely a place for boarding airport buses or limousines, but replicates some of the passenger service facilities to be found at the airport. At a city terminal of this type there are two principal groups of passenger-handling operations: the departure operations and the arrival operations. Departure operations are concerned with checking in passengers for flights, boarding them on coaches and dispatching them to the airport. Arrival operations relate to the disembarkation of passengers from coaches which have come from the airport, baggage claim and the dispersal of passengers from the terminal into the city. In designing a new terminal, or extending an existing one, it is necessary to determine the extent and nature of the facilities required by these operations, given traffic forecasts. We were supplied with such forecasts by the responsible department of our company—British European Airways—and were neither responsible for the methods by which they were derived nor for their accuracy. I do not therefore propose to discuss them. We shall consider the forecasting problem later.

The three physically different queueing problems which we shall consider today could be formulated in terms of one well-known model. Some bending and twisting of the model was found necessary to make it fit the conditions of the three problems, but it was preferable to distort the model rather than to distort the problems. The model used is defined in the customary manner as follows:

1. Input:

Customers arrive to demand service in a Poisson stream such that if the elapsed time between consecutive arrivals averages *a* minutes, its distribution $A(t)$ is given by

$$dA(t) = \frac{1}{a} \exp\left(-\frac{t}{a}\right) dt$$

2. Queue Discipline:

The waiting line (if any) is orderly, and the first customer to arrive is the first to obtain service.

3. Service-times:

The service-time per customer v has a probability distribution $dB(v)$ with a mean of b minutes.

4. Service-facilities:

There are s servers, any one of which can serve any customer (i.e., there are no specialized service facilities).

This is known as a multiple-server queueing model with random input and generalized service-times. It has been widely discussed in the literature and in two special cases charts are available which relate the customer waiting time w—which is the elapsed time from his arrival to his obtaining service—to the number of servers, and the average service-time and interarrival interval. These cases correspond to an exponential service-time distribution

$$dB(v) = \frac{1}{b} \exp\left(-\frac{v}{b}\right) dv$$

and a constant service-time

$$v = b = \text{constant}$$

Charts for these cases are given in the paper by Molina. However, in the practical situations described in this paper, neither of these models was exactly applicable, because in every instance it was found that the service-time distribution had the mathematical form

$$dB(v) = \frac{1}{(k-1)!} \exp\left(-\frac{v}{b}\right)\left(\frac{v}{b}\right)^{k-1}\frac{dv}{b}$$

This is known as the gamma (or Pearson type III) distribution. There are formulas available for this case too, but they are extremely complicated. However, approximations can be used in certain cases and fortunately the problems encountered in planning the terminal were of these types.

The first problem we may call, for short, the set-down curb problem. Observation showed that passengers traveled to the existing terminal in the following ways (figures are illustrative):

by taxi 75 %   by subway 9 %   other means 2 %
by car   12 %   by bus   2 %

It was intended that passengers arriving at the new terminal by taxi or car would be unloaded on a set-down curb outside a departure concourse; all others would enter the terminal area by foot and would by-pass this curb. Each taxi or car entering the terminal area from the street would drive up to the set-down curb via an approach road. If a vehicle were to find all curb-space occupied by preceding vehicles, it would wait in the approach road for a free space; then it would unload its passengers and their baggage on the curb, and drive away—taxis out of the terminal, cars to a park. The problem was to estimate the number of curb spaces required to ensure an almost total absence of waiting vehicles in the approach road. (The presence of stationary vehicles in the narrow approach road for any appreciable period could halt the operations of the terminal.)

A survey of the setting-down operation at the existing city terminal, established the following facts:

1. The average number of vehicles arriving per hour could be predicted from the scheduled number of flight-coach departures from the terminal during next two hours.

2. The vehicles, although arriving at this average rate, did so in a random—or Poisson—stream.

3. The curb-occupancy of a vehicle—the "service-time"—was obtained as a probability distribution later identified as a gamma distribution, with mean (say) 1 min and standard deviation 0.8 min.

4. The constriction of the approach road enforced a rule of "first come, first served" on the taxis.

This information sufficed to calculate the number of curb spaces, which would be needed to keep the queue waiting vehicles as small as desired, during the critical busy hours of future years.

An approximate method of calculation was used to overcome the difficulties associated with gamma service-time distributions. This made use of the relationship

$$Q = \frac{(1 + V^2)}{2} Q_e$$

where Q_e is the average queue-length if the service-time distribution were exponential with mean b, Q is the average queue-length if the service-time had a generalized distribution of mean b, and V is the coefficient of variation of this distribution.

This formula is exactly true when there is only one server, irrespective of the nature of $dB(v)$. Some simulation exercises which we had previously carried out on a related problem (Ref. 5) had shown us that this relationship was nearly applicable for many servers if the service-times had a common gamma distribution. Existing charts show that when service times are constant, so that $V = 0$ for example, the relationship

$$Q_c = \tfrac{1}{2} Q_e$$

is roughly true when the server-utilization U, defined as

$$U = \frac{b}{as} \times 100 \%$$

is not low in relation to the number of servers. This is a useful rule-of-thumb. Now, in fact, the coefficient of variation of the observed gamma service-time distribution was 0.8 which is not far from that of an exponential distribution (unity) and we used the approximation without hesitation. It was then only necessary to decide that the average queue length during the busy hour of the day should not exceed say 0.2 vehicles to work backward to the number of curb spaces required. This then was one application of queueing theory in the design of the terminal.

The next problem could be formulated in the first instance in much the same way as the set-down curb problem. Consequently, many of the principles and criteria which we adopted for its solution were identical and I need not make more than passing reference to them here. It is preferable to focus attention upon the importance of a different criterion to that of waiting-time or queueing-time distribution. The process in question concerns the checking-in of passengers for their flights. In the new terminal, as in the existing one, a passenger entering the terminal would have to check in for his flight at any one of a number of check-in counters, there being no segregation of passengers into flights at this stage. This constitutes a queueing process at first sight not unlike the last, but in fact there are some subtle differencs.

The input to the check-in process consists not of individual passengers but of passenger-groups—that is persons traveling together. A passenger-group may consist of 1, 2, 3, 4, or more persons. The service time for a group was found to be very nearly exponentially distributed, although in fact it had a gamma form.

Observation at the existing terminal showed that, considered as an undifferentiated mass, passenger-groups arrived again in a random stream at an average rate related to the number of scheduled coach departures during the next two hours. But considering the passengers flight by flight it was discovered that those passengers reporting for any one specified flight did not arrive in this way. First of all there was an upper limit on the number who could check in for a single flight, and second they had a deadline to meet in that they must arrive before coach departure. Now it is clear that if a passenger were to arrive so late that even though he obtained immediate service he could not be checked in before coach departure, he would have no legitimate cause for complaint. But if he arrived say 5 min before coach departure and missed his coach because he had to wait 4 min before being attended to, he might feel some justifiable annoyance. The problem to be solved here then was this: how many counters should be provided to ensure that the chance of a passenger who arrived before coach departure missing his coach would be small?

First of all, the time for which any arriving group would have to wait before reaching a check-in counter was determined solely by the overall queueing process considering all arrivals as equal, regardless of flight. As outlined in the previous example, it was possible to obtain the quantity

$$q(u, s) \, du$$

the probability of a waiting time (queueing plus service) of u minutes when there are s servers and the traffic in Erlangs is b/a.

Now observation showed that the passenger arrival pattern for any one given flight was given by an identifiable probability that any one group would arrive during the interval $(t, t - dt)$ minutes before coach departure

$$\lambda(t)\, dt$$

Thus the probability of this group missing the coach would be

$$\beta = \int_0^\infty \lambda(t) Q(t, s)\, dt$$

where

$$Q(t, s) = \int_t^\infty q(u, s)\, du$$

Now if N groups report to the terminal for this flight, the average number missing the coach will be $N\beta$. It was necessary to adjust s, the number of servers, so that this was very small (say 0.001). It is clear that this could be done without undue difficulty by careful computation.

The result of making this calculation was to decide that unless an unacceptably large number of check-in counters were to be provided, no group should be permitted to arrive—that is, join the queue—at the check-in counters later than 5 min before coach departure. Groups arriving later than this time should be directed to a special, "gate-type" check-in counter dealing only with last-minute arrivals. This analysis had a profound bearing on the layout of the departure concourse of the buildings as well as providing estimates of the numbers of check-in counters required. (Ref. 5 gives more details on this part of the project.)

The last of the three problems which I shall talk about differed again from each of the others in having special criteria—or *measures-of-effectiveness*—of its own. What it had in common with them was the multiserver model previously described. It concerns the parking of cars at the terminal.

A certain proportion of outgoing passengers would be brought to the new terminal in private cars by friends. After putting the passengers off on the curb outside the departure hall, the owner would drive his car into a short-term car park where he would leave it while he rejoined his friends and saw them off on the airport coach. He would then return to his car and drive it away. Similarly, a certain proportion of incoming passengers would be met at the terminal in private cars by friends. These friends too would wish to park their cars while waiting in the terminal.

Now observation at the existing terminal again sufficed to establish the length of time for which people parked their cars, and also the rate of vehicles seeking to park related to the number of incoming and outgoing flights through the terminal during the next two hours. Once more it was found that:

1.   Vehicle arrivals at the park during the busy hours were Poisson at a fairly steady average rate.

2.   The distribution of parking—that is, service-times—was of the gamma type.

It is clear, however, that there could be no waiting for parking spaces, either in the existing or in the new terminal. Any driver arriving to find the park full would have to park outside the terminal area on a side street. Thus, there was no question of determining the number of parking spaces needed to keep an average waiting-time below a specified figure. The problem was to determine the parking capacity which would ensure that only a small fraction of all arrivals would find the park full and be turned away.

This is an easier problem than to determine the average waiting-time, as the probability of all servers being occupied in a multiple-server queueing process with random inputs is independent of the shape of the service-time distribution and we were not therefore bedeviled by the service-times having a gamma distribution (see Ref. 5).

In fact, it is

$$P_N = \frac{1}{N!}\exp{(-b/a)(b/a)^N}$$

where the park can accommodate N vehicles. This is easily evaluated, and it was only necessary to determine N so that P_N would be small, say less than 1 in 50. This simple, but possibly not well known, calculation sufficed to provide estimates of parking capacity.

This was all ten years ago. The terminal has since been built and put into operation. Airline passenger traffic has continued to grow at a much faster rate than in 1958 anyone was prepared to predict; consequently, it is quite improbable that the terminal will remain useful for as long as was anticipated. The errors in the forecasting of the parameters of the queueing problems have proven much greater—as they often do—and have had more effect upon the validity of the calculations than any of the over-simplifications of the model and its approximations could possibly have had—short of total inappropriateness. But these old problems and the way we set about solving them pose another question: if we had a similar job to do now, would we do it the same way? The answer is no. The need to make approximations to keep the computations simple is no longer strong. In Air Canada, for example, we have computer programs for calculating waiting-time and queueing-time distributions numerically for a wide variety of service-time distributions corresponding to the multiserver model $M/G/c$: (FIFO/L) which was encountered in 1958. It is a matter of minutes to obtain results once the data are fed in to the computer. In the case of more complicated queueing processes than those we have discussed, undoubtedly one would be inclined to obtain results by Monte Carlo simulation methods. Computers seem to have made all the difference. But perhaps not all: the third problem we should still solve without their help.

3. SIMPLIFICATION, REALITY AND PRACTICE

The three queueing problems which I described in the last section were not, of course, the only queueing problems, or the only type of problems, encountered in the advance planning of the West London Air Terminal. The whole series of interrelated studies was quite conventional in some ways: the scientific method was used; the work was done by a team (a mathematician, an anthropologist, an economist and two assistants); both information and operational subsystems were looked at; and finally the results were presented as advice to management (who accepted them). This was operational research in the classic sense. A fuller account which, I hope, supports this somewhat immodest assertion is presented in my book *Applied Queueing Theory*.

In other ways these queueing problems were not at all typical of much operational research. In the first place the problems and the processes to which they related were *well-structured*. That is to say the significant factors that must be considered in solving the problems, and the relations between them, were easily recognized because the problems were generally of a known type. In applying the scientific method to their solution, the tasks of data-collection and data-reduction, which are so often tedious due to the initial accumulation of subsequently irrelevant material, were simplified. When a problem, or process, is well-structured this is invariably so. Furthermore, to find an appropriate model was not difficult: all that was needed was to identify certain probability distributions implied by the observations of arrivals and service-operations. It is true that when this was done we found that the appropriate models (in two cases) were not ones for which formulas suitable for rapid manual computations existed; but it was not difficult to find solutions which, though approximate, were sufficiently accurate for practical proposes. That is to say *operational solutions* could be fairly easily calculated.

While it is common enough that analytical techniques appropriate to a problem situation should exist, it is, in my experience, somewhat rare to find usable models and formulas handy, and well-structured problems and processes are infrequently encountered—at least in the present state of our knowledge of the systems and processes to be found in organizations. Most are ill-structured, or at best partly-structured. As a general rule, therefore, an operational research man ought to have some facility in devising concepts and models. He must be able to exercise some creative imagination in stage three of the scientific method. Possession of this facility to any marked degree is not common. It is not surprising then that operational research workers (even at times the most distinguished) exhibit a tendency to look for problems in well-structured situations for which familiar models and methods of calculation are readily available. The principal danger of this tendency, which has been often noted, is that problems are not tackled because of their importance but because of their amenability. This is, however, to digress. The point I wish to make here is that operational research is something more than the application of ready-made models and formulas. It is not queueing-theory only. Above all, it is not merely the application of standard linear-programming algorithms. This brings me to another aspect of the three queueing problems.

Objective Functions and Optimization

When I published a book some two or three years ago which contained a fairly extensive account of the queueing problems encountered at the West London Air Terminal, one reviewer complained that, properly speaking, they were not examples of operational research at all because in no instance was there any consideration of an *optimal solution*. In particular, he was disturbed that no *objective function* had been set up to measure the sum of the cost of operating and owning, for example, a check-in system on the one hand and the cost of delays to passengers served by it on the other. He implied that the best system would have been that which would minimize that value of this objective function. I am happy to have the opportunity to declare this view ingenuous. The mistake was a natural one for, as we shall see presently, to the classical list of characteristics of operational research has been added in recent years:

5. The solution, or course recommended, or system or procedure designed shall be the *optimal* one in terms of the value of some objective function.

In other words, operational research is to do with *optimization* of system performance. I do not quarrel with this. It is in fact a basic principle: but it must not be applied unthinkingly, and sometimes in practice it cannot be applied at all.

Let us consider the taxi set-down problem. The cost of a curb-space for a taxi, or private car, could be easily established to be sure. Say it was $\$c_1$ per year. If there were k spaces the total cost would be $\$kc_1$ per year. Now what would be the cost of delays to taxis, cars, and their passengers? At any traffic loading k spaces would result in some average delay per vehicle—this is oversimplifying for sake of illustration—of say $w(k)$ min. What would be the cost of a minute's delay? The cost to the taxi and driver could, as a first approximation, be set at the marginal cost of waiting-time as registered on the taximeter: say $\$c_2$ per min. Now what about the value of a minute's delay to a passenger? This was not known and could not be established easily or at all. To attempt to obtain a reliable estimate would have cost a great deal of money—more, unquestionably, than the cost of providing extra spaces or of delays to a few passengers—and what is worse taken such a long time that the results would not have been available in time to be used. Operational research is concerned in situations such as this with obtaining practical, useful solutions to problems of an executive type—i.e., operational solutions—not with formal elegance or academic approbation. In operational research the practical constraints and pressures are always important parts of the problem situations: or, where this is forgotten—as it sometimes is, even by management—let us say they *should* be.

Similarly, in the case of the passenger check-in process, the objective function was not, and should not have been, simply a matter of the operating and ownership costs and the value of the delays to passengers (even if the latter could again have been estimated, which it could *not*). The check-in system, like all systems in the universe, was only a subsystem of some larger system. It was not so important to minimize a cost-based objective function within the *system boundary* of the check-in system as to ensure that passengers would not have to endure such long delays in the queues before check-in counters that they would divert their custom to other airlines or other means of transport, or would simply cut down upon their travel altogether. To have derived a solution from optimization of a financial objective function it would have been necessary to expand the system boundary to include the presumed effect of queueing-time at check-in counters upon traffic volumes. With some time-lag (and now the real extent of the complications becomes apparent) reduced rates of traffic growth—or even of base volume—might reasonably be expected to follow unbearably long queueing-delays. But the precise nature of this relationship was not, could not and in fact need not be known. The management decision to maintain a low level of passenger delays, undoubtedly below whatever critical threshold might exist, effectively contracted the system boundary so that is became *uncoupled* from potentially interacting subsystems and processes. Once this was done it was unnecessary to construct and optimize a cost-based objective function; for either the result of the optimization, subject to the constraint that the average queueing-time should not exceed w_q min., would merely be to provide the minimum number of counters needed to support such a maximum delay; or to provide a standard or service much higher than would be of any use to passengers at greater expense to the airline and eventually, and incalculably, to the passengers themselves. For the system boundary had been drawn to exclude the relationship between the cost of check-in service and the fare-level necessary to sustain it. As Fred Allen once said, "Most things are more complicated than they seem to most people."

Another way of putting this is that once the path of system optimization is followed, it is commonly necessary to expand the system boundary to avoid misleading suboptimization. Suboptimization is to optimize performance of a subsystem with respect to an objective function of internal variables when significant relationships between these and some external variables have been ignored. Charles Hitch pointed out the dangers of doing this some years ago and indeed they are real enough. However, as we have seen, indefinite expansion of the system boundary may result only too often in no problem ever being solved. Practical compromise is necessary. It is such exercise of judgement by operational research workers—and by management—that makes the effective practice of operational research so much of an art as opposed to a science. I can try to sum up this discussion in making three points:

1. Definition of the system boundary—what elements, factors and

relationships are included—is of the first importance. If too much is excluded, wrong conclusions may be reached; if the boundary encompasses too much, no conclusions may be reached at all.

2. A judicious choice of constraints may serve to define a manageable subsystem which may be analyzed and optimized to a degree adequate for practical purposes.

3. The cost of obtaining the information, and developing models, needed to perform an optimization should be included in any cost-based objective function. This is, of course, and accepted practice in devising statistical decision schemes in which the cost of acquiring sample data must be set against the cost and risk of errors.

Models of Subsystems of Organizations

One of the most striking characteristics of organizations—as indeed of many natural systems—is that they change state as time passes. In other words, they *evolve* or are dynamic. Now it is remarkably difficult to study, analyze, make hypotheses about, and predict the behavior of dynamic systems unless they are cyclic—as for example the solar system very nearly is. If they are truly evolutionary they are difficult to model. Consequently, in operational research one must resort to numerous stratagems of a technical nature to be able to construct usable models of organizational sub-systems. For example, a queueing-process (or subsystem) such as the check-in system at an airport is an evolutionary system: people change, procedures drift or adapt informally to changing conditions, and even at the level of analysis represented by "classical" queueing theory undergo dynamic change. As the servers become more expert, their rate of work not only increases but becomes adaptive in relation to the amount of work to be done. The average volume of traffic tends to increase during the period of a schedule. And other changes occur. Consequently, any classical queueing model tends to relate only to some idealized, simplified view of reality. Organic systems are inevitably more complex than inorganic and—if I may say so— biology is much more complex a study than physics. Indeed, to digress for a moment, many of us suspect that the successes and attractions of physics are due to its apparent, *relative* simplicity.

If we assume, however, that the classical formulations of models of queueing processes and stochastic service systems (which ignore parameter drifts and interactions between service-rate and queue-sizes) are valid approximations to real processes in the short term, we face considerable technical problems. For example, let us suppose that the input rate to a queueing system is constant: then the transient phenomena are, except in the simplest cases, not only extremely intractable as problems in mathematical analysis but at higher traffic intensities are in practice very prolonged, as simulation studies have demonstrated. Consequently, in any real-world situation, where it is unusual to find a constant traffic intensity for very long, most queueing-processes hardly ever emerge from a continuously shifting transient condition.

Steady-state solutions are therefore no more than convenient approximations to approximations. The solutions—if I dare now use such an appellation—of the problems discussed in the previous lecture were no more than this. The concept of a steady-state or equilibrium solution is a *device* by which guidelines to understanding and decision may be obtained. So long as an operational research scientist realizes this he cannot go far wrong, and with moderate ingenuity he may be able to guard against gross error. It is a common tendency, however, for models based upon idealized contractions of the boundaries of real systems to be seen as reality. This can both be misleading and lead to conflicts of understanding between analysts on the one hand and "practical" men on the other. I wish to warn against it.

The way around this difficulty is one familiar to engineers who have for many decades been concerned with the design of man-made, *physical* systems. Engineers do not ignore the precepts of physics and mechanics simply because no such things as ideal gases and ideal metals exist. They use scientific knowledge but allow for error and its consequences in theis designs, and in doing so balance costs and risks. This is what an operational research worker must do when applying the results of his analysis of systems and processes to their improvement, or the design of their replacements. As I have said earlier, true operational research must be as close in concept to engineering as to science. Consequently, we must concern ourselves greatly with the *robustness* of our models and their sensitivity to errors in their parameters. It is commonplace in telephone engineering, for example, to provide for at least a 10 percent error in traffic forecasts. I have been involved in projects where a 30 percent error in forecasts of parameter values had to be allowed for.

Now this has implications for model-construction. The construction of symbolic models of systems of all sorts is a fascinating business. It is attractive to persons of a speculative turn of mind—as indeed theoretical physics seems to be to many people whose practical exposure to physical phenomena in experimental laboratories is negligible. Endless discussions of the finer points of model formulation are possible. But in practical operational research they are frequently profitless. The variations between the solutions to practical problems suggested by different models are quite often smaller than those resulting from errors in the sample estimates of parameter values. If time is available and money is to be spent, it may often be more useful to collect better data than to devise more refined models. The decision as to the optimal distribution of investment of time and money between data-collection (and processing for parameter estimation) on the one hand and model-formulation (testing and perhaps programming) on the other must itself be made rationally.

Research and development in the planning of research and development is a proper subject for operational research. This is an attractive theme, but is in the nature of a digression which I have not time to pursue here. Let us in conclusion return to the modeling of dynamic systems.

Many of the stratagems currently employed in operational research are directed at devising *static* models of *dynamic* systems. Statistical equilibrium is a stratagem for reducing stochastic processes (which are dynamic) to problems in probability distributions (which are static), by means of time-suppression. "In the long run," said J. M. Keynes, who might almost have been commenting upon ergodicity in evolutionary systems, "we are all dead." It is a useful device, and necessary, but nevertheless it is a *device*. Similarly, on a lower level, deterministic models are often a simplification of probabilistic phenomena. In the next section we shall consider some examples.

4. UNSTRUCTURED DECISION PROBLEMS

The operational situations discussed in the second lecture have been described as *well-structured*. That is, the behavior of such queueing systems has been studied quite extensively, and both the principal variables to the considered and the relations between them are fairly well known. Reasonably appropriate models at the third level of conceptual sophistication have been—or may approximately be—defined. Stochastic models with time-suppression (i.e., the "steady-state" fiction) have been much applied with results that are acceptable operationally. I have asserted that most—or anyway a substantial—proportion, of the problem situations encountered in the functioning of any Organization are not well-structured in this sense. Commonly situations are to be observed in which no one—not even the most assured man-of-experience—really knows what the structure of the problem is, or can identify the significant factors with any rationality, or has any clear vision of the relationships between those that he can identify. One way of resolving such situations is by the application of conventional judgements. The answers may not be correct, but at least everyone knows how they were arrived at. This is characteristic of, for example, much cost-accounting.

In this section I shall draw upon some work which I did between 1958 and 1962 to illustrate the development of a structure for an hitherto ill-structured problem and the production of a mathematical model. It is also an example of an *optimization* problem in business, occurring in recurrent, decision problems at the middle-management level—one level higher, perhaps, than that where operational problems are customarily resolved. The technical exposition will be brief because it is readily accessible elsewhere (Ref. 7 and 8) and because it is no longer the latest word on the subject (see Ref. 9). It concerns advertising. Its nature is as follows. Suppose that an organization decides to spend some money during a certain period and in a certain region to promote its products. Then someone must decide how this sum of money should be allocated among print media, television, radio and outside advertising to the best effect. Now the ultimately desirable information required here would be the sums of money to be allocated to each medium; in other words, the complete solution. Unfortunately, although these "system objectives" are defined, the problem is not fully-structured because no models and algorithms exist for producing solutions to such problems. Rather oddly, however, the type of data that must be processed in some way to arrive at a rational solution is quite well appreciated. It includes statistics on *readership* and *reach* of print media, radio and TV audiences, the noting values of different advertisements, Starch ratings, audience ratings, market descriptions, in terms of demographic and socioeconomic variables and so on. What is missing is a conventional or established means of operating on these data to produce answers of the type required. Consequently, the "rule-book" takes the form of human judgement, informed and intelligent no doubt, but variable from judge to judge. It is certainly not based on a universally accepted and dispassionately validated mathematical model or upon conventional, agreed rules.

Defining the Boundary

The design of an advertising schedule that is in some sense optimal was, and still is, a formidable problem. My colleagues and I, when faced with it, began by constraining it to manageable proportions. We took the following general view of the matter.

An advertising compaign is usually carried out either to persuade some portion of the population to take some definite course of action, or to create and maintain an idea or an attitude in their minds. Whatever the objective may be, the essence of advertising is that a message must be conveyed from the advertiser to some class of people called the *target population*. The message must be expressed in coded form; words, pictures, ideograms and so on. The method of expression is called the *format*. The message, in whatever format, needs a *carrier*— that is a medium through which it can be transmitted to the target group. This medium might be the press, television, boardings, or any one of many other means of mass communication.

We assumed that market surveys had shown that the people most likely and able to make use of an excursion air-fare constituted a certain segment of the population defined in terms of age, occupation location and income. These people constituted the potential market and therefore the known target population. Furthermore, our company had already had to discover what message it should convey to the target group to appeal to it and encourage the people in the group to travel. It had carried out motivational studies to find this out and had decided to convey a certain message. Lastly, the company's advertising agency would have already produced a format for the message. The problem was solely one of drawing up a media schedule which would be an efficient carrier. After some consideration, in which experience and judgement must have come into play, the media specialists had, we assumed, decided that the campaign must be of short duration, intensive and intended to make a considerable, heavy impact on the target population.

All of this amounted to a very considerable contraction of the system boundary. For example the interaction, possibly important, between the format and the media schedule was excluded. Nevertheless, it was too big. Shortage of data and reliable research results for other media decided us to consider newspapers and magazines only. (This is equivalent to the effects of lack of knowledge of the effect of queueing delays on traffic volume in the queueing examples: to have obtained it would have been ruinous.) We believed it possible to assume that the sum of money available for buying advertising space in newspapers and magazines was fixed and then to proceed to allocate it amongst the publications available in an optimal manner. When a method for doing this had been developed and tested, similarly restricted research could be carried out for television and so on. Eventually a synthesis could be attempted. Others are investigating these extensions today.

The Model Formulation

Let me begin with some definitions.

1. The size of an advertisement in any publication is defined as the proportion of the total area of a whole page in that publication which it occupies. It is denoted by p (or z^2 for reasons which will become apparent).

2. A series of advertisements of size $p_i = z_i^2$ in any medium M_i, appearing one at a time in each of n issues will be called a multiple sequence of dimension (z_i^2, n). The ultimate coverage of such a multiple sequence is defined as the expected proportion of persons in the target population P who will see at least one advertisement of the series, and is denoted by $G(z_i, n)$. Similarly the ultimate impact, or average number of advertisements in the sequence seen by a member of P is denoted by $F(z_i, n)$.

3. A campaign consisting of k multiple sequences in k media $M_1, M_2 \cdots M_k$ where the dimension of the sequence in M_i is (z_i^2, n_i), $i = 1, 2 \cdots k$, will generate an ultimate impact $F(z_i, n_i; k)$ and an ultimate coverage $G(z_i, n_i; k)$.

The next step then was to identify the forms of F and G. We had to make a number of assumptions. The first of these was that the size and attention value of an advertisement are related by the square-root rule:

1. An advertisement of size $p = z^2$ in any publication is seen by a proportion z_i of the readers of that publication.

This assumption was based upon some experimental work carried out by London Press Exchange and implies that an advertisement of size 1, that is a whole page, is seen by every reader of a publication. The second assumption was that

2. The proportion of the target population who read both media

M_i and M_j, a_{ij} say, is the product of the proportions a_i and a_j, respectively, i.e., $a_{ij} = a_i a_j$

This assumption is an approximation to the truth, which may not always be valid. But in press and magazine advertising, in Great Britain at least, it was a good approximation. The third assumption was that

3. The chance that a reader will see an advertisement in M_i is independent of whether or not he has seen an advertisement with the same format and content in M_i previously or in another medium M_j.

We also assumed (though this may easily be relaxed) that the cost of an advertisement was proportional to its size, p.

These assumptions were the basis of the model. Every one is an over-simplification of reality as most assumptions in scientific work have to be. But these were experimental and statistical bases for them. We were then able to proceed, for if an insertion of size $p_i = z_i^2$ is made the basis of a sequence of n_i repetitions in medium M_i and the readership of M_i is a proportion a_i of the target population, then

1. the ultimate impact of 1 insertion in M_i is $a_i z_i$,
2. the ultimate coverage of 1 insertion in M_i is $a_i z_i$,
3. the cost of 1 insertion in M_i is $c_i z_i^2$, where c_i is the cost of a whole page.

Furthermore,

4. the ultimate impact of the sequence in M_i is $n_i a_i z_i$,
5. the ultimate coverage of the sequence in M_i is $a_i[1 - (1 - z_i)^{n_i}]$,
6. the cost of the sequence in M_i is $n_i c_i z_i^2$.

The third basic assumption permitted us to derive from these results:

$$F(z_i, n_i; k) = \sum_{i=1}^{k} n_i a_i z_i$$

$$G(z_i, n_i; k) = 1 - \prod_{i=1}^{k} \{1 - a_i[1 - (1 - z_i)^{n_i}]\}$$

$$C(z_i, n_i; k) = \sum_{i=1}^{k} n_i c_i z_i^2$$

While many combinations of problems might have been devised, two were of primary interest. The first of these, which we called the *impact maximization problem*, may be stated as follows: Let $F(p_i, n_i; k)$ be the average impact on the target population of r_i repetitions of an advertisement of size p_i in the K media M_i ($i = 1, 2, \ldots, k$), and T be the total cost. Find the values of n_i and p_i which maximize F for this cost. The second, the *coverage maximization problem*, is if $G(p_i, n_i; k)$ is the coverage of the target population achieved by r_i repetitions of an advertisement of size p_i in the k media M_i ($i = 1, 2, \ldots, k$), to maximize G subject to the cost equaling T.

I do not propose to work through all of the mathematics here. Both problems can be solved by the method of undetermined multipliers (Ref. 7). For interest, however, here are the results for the coverage maximization problem:

1. No n_i should exceed unity.

2. The values of z_i which provide maximum coverage are given by the k equations

$$\frac{Ha_i}{1 - a_i z_i} = C_i z_i \sum_{j=1}^{k} z_j \frac{a_j}{1 - a_j z_j} \qquad (i = 1, 2, \ldots, k)$$

Except that if this requires $z_i > 1$ for any i, then that z_1 shall be put equal to 1 and the corresponding media be removed from the calculation which will be repeated for the remaining media and money.

Computers and Operational Research

One of the principal features of the solutions we obtained to these problems is that they are not expressible in purely algebraic forms. The equations embodying the solution to a coverage maximization problem, for example, can only be solved by tedious, iterative numerical methods. This requires the use of computers if solutions are to be readily obtained. Such is the position in many operational research studies though—as the queueing examples have shown—not by any means all. It is in my opinion doubtful whether operational research could have developed in the way it has had computers not become commercially and economically available in the late 1950's. From the point of view of operational research, elegant formal solutions are in themselves valueless; it is essential to compute numerical solutions which can be used.

Now advertising campaigns must be planned, as many management decisions must be taken, recurrently. That is, problems of identical form frequently recur. Consequently, it is worthwhile to program a computer to solve these recurrent, structured problems. But in practical terms this means that it is not enough to devise a method of solution or even a computerized algorithm for it. An entire system, and a set of procedures, must be devised to ensure that

1. The input data are always up to date and on file.
2. The users know how to frame their problems and are given a means to do so.
3. The computer operators know how to run the program.
4. The users know how to interpret the results.

This necessitates a broad approach, similar to that used in the original operational research studies in Britain before and during World War II. In particular, computer systems analysts must be integrated into operational research teams.

Computers should not, however, be regarded as a subsitute for thought. I am afraid that they sometimes are so regarded on the general but absurd principle that if sufficient data are collected, sorted, reduced, and tabulated in a large enough variety of ways, the problems of planning and decision will disappear. "When in doubt," it has been cynically remarked, "simulate!" One can often do better by more carefully reformulating the problems. For example, the approach to media scheduling which I have described resulted in equations requiring extensive computation. Subsequently, my colleague, C. J. Taylor, by adopting a marginal cost viewpoint which I had previously tried and foolishly discarded (following some trial calculations made for me by W. S. Chatham) and by making some ingenious approximations, managed at one stroke to surmount this difficulty. He succeeded, in fact, in devising a computing procedure for obtaining these answers which is so clear and simple (always a sign of real success) that no computer is needed at all. He subsequently wrote to me to say that as a result of an extension of his marginal cost analysis, he had been able to deduce an "optimal size decision rule" which yields for all press media the absolutely most effective size of advertisement which should be employed, given empirical data concerning the attention values of different sizes of advertisements and cost data. With the help of this rule a very general formulation of the problem of choosing the advertisement sizes, frequencies of insertion and press media which will maximize "response" for a given expenditure has been reduced to a logically straightforward computational procedure. Although Taylor claimed that this could be followed on a desk calculator fairly quickly, it is my opinion that the procedure is sufficiently elaborate to justify the use of electronic computing equipment.

Conclusion

It may be disappointing to some that although in this section I have presented an optimization problem I have not talked about *linear programming*. This was deliberate. So much has been written about that technique that it would be superfluous to add to it. However, to assuage this disappointment I am pleased to say that some advertising problems may be appropriately formulated in terms of linear programming and so solved.

5. MATHEMATICS AND FORECASTING

Writers on the art of management commonly assert that the top managers of any organization should not involve themselves in unnecessary details of its work. Few of these writers, I suspect would admit as matters of valid, detailed concern to any top management either the solution of queueing problems in the technical design of airport terminals or the construction of optimal media schedules. It is a complaint often made of—and indeed, by—industrial engineers that their years of training and their store of expertise are devoted more often than not to the improvement of the lowest-level and least consequential work-tasks in their companies. I do not deny that similar charges are made of—and again, by—O. R. workers from time to time; nor are they completely without justification. I cannot promise to provide further examples of operational research studies that are not trivial from a mathematical point of view, but I can and intend to describe two today that are very closely related to the genuine concerns of top management. In this way I hope to convince you that O. R. is not just a form of mathematical work study.

The writers on the nature of management seem to agree that one of the principal tasks of the top management of any organization is to formulate the policy defining the goals which it will pursue and the limits within which the organization must conduct its operations. When such a policy exists and is clearly understood one major task of the next level of management (or of the same acting in different *rôles*) is to develop in some detail the precise strategic plan that appears to implement the policy best. That is, these managers must make strategic planning *decisions*. The time-scale of strategic plans is customarily extensive as they relate above all to major issues: the design, selection, and procurement of important facilities and the recruitment, development and training of managerial and specialist personnel. The elapsed time from taking a decision on such matters to its implementation must usually be long. If a plan is needed for implementation in any one year, the plan itself must be drawn up initially several years ahead. For practical purposes then, strategic planning and long-range planning are synonymous.

As the time approaches for the implementation of a strategic plan, the executive must turn his attention to the details of implementation in the light of current knowledge: that is, he must begin tactical, or short-range, planning. In doing this, he will be constrained by the limits created by previous strategic plans: thus his range of choice of tactical plans is restricted. For the most part, tactical plans relate to the precise determination of what to do with the equipment and personnel that will be available, in terms of deployment and assignment of tasks. The elapsed time to implementation is relatively short.

On the day of any particular operation, of course, operating management will take control decisions to adjust the existing tactical plan so that it conforms to the current realities of the situation. But operating decisions have a limited time-scale: they are decisions made at a certain time, on the basis of information available at that time, to be implemented immediately or very soon, and which will in general not have far-reaching effects. Planning decisions are not like this. They are decisions taken at one point in time, on the basis of forecasts then available of the situation at some future time, which are to take effect at the future time.

Forecasting, then, is essential for planning and sober planning and is —or, if it is not, should be—a major and constant concern of top managements. The design of effective tools for forecasting, and perhaps even more the design of organized systems that will produce valid forecasts and deliver them to management, would seem to be valid subjects for the O. R. workers in any rational organization to interest themselves in. My colleagues and I in Air Canada have been concerning ourselves with such matters now for about seven years. Not I hasten to add, that our concern has been exclusive or by any means continuous: on the contrary our work in this field has gone forward in fits and starts. For the most part the periods of quiescence were those in which the existing planning staffs were supposed to be familiarizing themselves with the new tools and testing them. They were so busy making forecasts and plans by hand that they could not plan to set aside sufficient time to learn what was new. This is a situation commonly encountered in the practice of operational research or any other variety of management science. That is, however, to digress. To return to forecasting: we have developed and implemented three principal mathematical methods for obtaining forecasts. Each relates to one particular type of planning and they are not much similar in any way. Let us begin by seeing why this is.

The great discriminant between long-range (strategic) and short-range (tactical) planning is, as the names imply, time-scale. As we have noted already the time-scale of strategic planning is extended. For example, the lead-times associated with strategic planning decisions in airlines are of the following order: five years may elapse between the time it is decided to erect a new airport terminal and the time is goes into operation; two to three years is a not uncommon lead-time for the delivery of an aircraft of a type already in production; for a new type a lead-time may be seven years. During such intervals of three to seven years structural or volume changes in the national and regional economies of the markets served, which greatly influence traffic growth, may occur. Measures of factors external to the boundary of the organization itself must therefore be major inputs to the forecasting process. Also, the possibility of desirable shifts of emphasis in policy during such an extended period may be anticipated; and the effects of such shifts upon the forecasts of the input parameters of the forecasting model must be assessed so that, in turn, the merits of these shifts may be evaluated. The forecasting techniques to be used for such purposes as these must— unless the results are to be obtained by magical incantation (as they often are)—necessarily be complex. We may conclude that an econometric forecasting model is a decided requirement.

The elapsed time from taking a tactical planning decision to its later implementation is relatively short: six months, perhaps, or at the most a year. Very often it is reasonable to assume that during such intervals no significant discontinuities in external influences upon the organization can occur and that no decided policy swings will develop within. Thus, currently identified patterns of development and growth may be assumed to continue more or less unchanged into the near future. In such cases some sound mathematical method of efficiently extrapolating the historical record is what is necessary; and if the rules for its use can be framed to signal the breakdown of the continuity assumption, it is likely to be also sufficient. Econometric analyses are not essential.

Long-Range Forecasting

First of all, I shall describe quite briefly an econometric model which we have developed at Air Canada since 1963 for use in long-range forecasting. It is called the Econometric Model for Marketing which, in short, is *EMMA*. The *EMMA* system is based on an econometric model which relates socio-economic (and policy) variables to air traffic with a view both to estimating elasticities and to forecasting traffic for three to ten years ahead. The traffic is measured in number of passengers and, at the discretion of the person using *EMMA*, may relate to an *Origin/Destination* (O/D) pair of cities and the flow between them or to the flow within a specified geographical boundary such as, for example, that of continental Canada and the United States. The calculation of estimates of elasticities and traffic are achieved by a series of computer programmes which basically carry out a correlation analysis, and feed forecasts of the socio-economic variables into a regression equation. The socio-economic variables are measures of external influences over which Air Canada can exercise no control, so that *EMMA* is first and foremost a system for processing externally generated data into the form of information such as rates of growth that is meaningful for internal planning decisions.

When we first commenced work upon the development of *EMMA* we had to decide which socio-economic variables should be included in the model and which policy variables should be allowed for. To decide which socio-economic variable to use, the members of the *EMMA* project team asked themselves and a great many others why should people travel. To each reason for traveling they attached socio-economic indicators which would reflect changes in travel for the reason. As far as possible the data collected related to the Origin or the Destination. To determine which policy variable to use, they asked themselves, and consulted market research surveys, to identify the aspects of travel which people consider when making a journey. As far as was possible they associated a measure of public attitude towards each aspect. This gave about 50 variables that could be used in forming a model.

The regression equation is a statement of cause-effect relationship and as such must be logical. Some of the data proposed in their unreduced state relate to a population mass, e.g., disposable income, while others refer to a person, e.g., retail price index and fares. It would be illogical to use combinations of these variables in a regression equation as they were. It is more logical to put all variables on a per capita basis so that they refer to a person. Consequently, *EMMA* relates income per head, production per head etc., and fares etc., to traffic per head. Traffic per capita is predicted from an equation. The coefficients of this equation are found by a regression analysis of historical data. The equation is in the form:

$$X = AY_1^{\alpha_1} Y_2^{\alpha_2} \cdots Z_1^{\beta_1} Z_2^{\beta_2} \cdots$$

X represents traffic per head. The Y terms represent the moving averaged socio-economic variables on a per capita basis. In the analysis the traffic per head variable for a particular month is taken with the socio-economic variable moving averaged over a period up to and including that month. This imposes a six-month lag on the socio-economic variables. The Z terms represent the policy variables and are related to traffic per head for the same month. The α's and β's are elasticities of traffic to the variable they are powering. They represent average elasticities for the period of the historical data. A is a constant.

To obtain predictions of traffic per head, forecast values of the socio-economic variables are fed into the formula. Values of the policy variables at the last month of the historical period are also used. Forecast values of the socio-economic variables are made from the historical data using trend prediction in conjunction with the exponential smoothing method of finding the trend level and its monthly rate of increase. In fact four further rates of increase are evaluated by adding an annual 1 % and 2 % to/from the exponentially smoothed rate. These five rates of increase are used to give five forecasts of the socio-economic variables (forecasts at five different levels of growth). Using forecasts at each of the five levels in the formula separately gives five different prediction of future traffic. If more than one socio-economic variable is being used to forecast traffic a check is made to see that the relationship between these variables that occurred in the historical data is being maintained in the forecasts.

The forecasts of traffic per head are then multiplied by forecast population; the forecast of the population being made externally to the system. A further program allows the user to adjust the forecasts as a result of policy changes or for changes in forecasts of the socio-economic variables. The noteworthy features of the *EMMA* system are:

1. It is more general than any particular application requires, but for any particular application it is not fully-structured. (Some human intervention necessitating the application of judgement takes place.)

Consequently, for any particular application it always turns out that many of the input data are redundant.

2. Apart from the few points at which human intervention occurs, it is programmable.

3. Input data cannot be collected in real-time or regularly, and are in fact generated externally in a completely uncontrollable manner. The data are extensive in variety and time-periods covered. (Historical records needed are voluminous.)

4. The system can deal with recurrent problems according to a predetermined schedule; and ad hoc problems only with relatively large delay owing to the need for special data collection and preliminary reduction.

Short-Range Forecasting

Every month the computer center of Air Canada's accounting department, situated at Winnipeg, compiles a matrix of the Origin/ Destination passenger flows for the previous month. These data are put to many uses. A magnetic tape containing the elements of the O/D matrix is transmitted to the Montreal computer center near to the head office. There it becomes the input to a medium-range (or short-range, depending upon one's personal viewpoint) mathematical forecasting model known as the *Trend-Projection System*. This system, which predicts passenger flows by month for all important O/D pairs up to two years ahead has two parts to it:

1. Projecting forward historical patterns of passenger flow.
2. Evaluating and modifying these projections in the light of current and expected future conditions.

The forward projection of historical patterns consists of a large amount of mechanical work, and in the "trend projection system" this is performed by a computer: the "rule-book" for producing information about the future from data from the past is programmable in this phase. The rule-book is based upon a particular type of moving average, the Exponentially-Weighted Moving Average (EWMA), which has the advantage that past data need not be retained. It is calculated in the following way:

New EWMA = Old EWMA + α (New Value − Old EWMA)

i.e.,

New EWMA = α (New Valve) + $(1 - \alpha)$(Old EWMA)

where α the smoothing constant, can take any value between 0 and 1. The EWMA has several advantages for a computer application. Its use requires very little preliminary analysis (merely calculating a starting value for the average). It takes up a minimum amount of storage in the computer. By varying α we can alter its speed of response to changes in passenger flow; and it is simple to introduce the new value to up-data the average each month. We shall see in the next section that the EWMA is in this instance a device for making a non-Markovian process look like a Markovian process. It manages to package the entire historical record into one figure carrying the time-label for the previous event.

The mathematical model of typical time-series of passenger traffic data assumes that a monthly figure may be synthesized from four components:

1. An average level.
2. A trend or growth component.
3. A seasonal pattern.
4. A random of irregular component.

The last of these is, by its random nature, not predictable, but the first three may be estimated and projected into the future. The process uses EWMA's to continually re-estimate and up-date the average, trend and seasonal pattern each month. The smoothing constants used are alpha (α), beta (β), and gamma (γ). To produce predictions the process uses the up-to-date estimates and projects them into the future. The more stable these components are, the more accurate are the projections produced. The difference between the predictions and the true values is a measure of the magnitude of the random variation present in a series.

The prediction process is like a "black box" into which is fed historical passenger data. The outputs are the predictions. We can control the process by altering alpha, beta, and gamma. To do this we need an early warning of changes which send the process out of control. To obtain such warnings of change, quality control techniques are used. Tests on selected series showed that the ratio of actual monthly passengers to the predicted value made in the previous month (A_p) was normally distributed with a mean of 1 and a variance σ^2 which differed from series to series.

In the control scheme every month when the actual volume is obtained as an output from the reporting system (also referred to in Note 1) as monitoring information, the ratio A_p is compared to the control limits $1 - 3\sigma$ and $1 + 3\sigma^*$. The chance of A_p being outside these limits if the process is under control is less than 1 in 300. Therefore, if A_p is outside the limits a cause must be found for the change in passenger flow. The nature of this cause, together with the duration of its effect, determine how alpha, beta, or gamma must be changed to bring the series back under control. These changes are effected by intervention on the part of market analysts, the users of the results. In a real sense the composite man/machine system is adaptive. We shall take this up in the next section.

6. PATTERNS OF OPERATIONAL RESEARCH

At the end of the fourth section I spoke of the incorporation of a mathematical model and method of calculation into an information/ decision-making *system*. Operational research projects increasingly, in my own experience, end up as exercises in the design and implementation of systems. That they may not necessarily do so is, however, illustrated by the queueing examples of the second section. We may, in fact, make a bold generalization and say that operational research work may be divided into two classes:

1. That which concerns the solution of one-off, *occasional* or nonrecurrent problems; or the evaluation of alternatives for an individual, nonrecurrent decision.
2. That which produces a method of solution for *recurrent*, or *regular*, problems which may be systematized by expressing the method of solution as a "rule-book."

It is, I believe, of interest to investigate further the conditions which give rise to operational research of each type and what the implications are for the practice of operational research in relation, say, to computer systems design and industrial engineering. In doing so we shall develop a simple framework for the classification of operational research programs. A *framework* is the simplest form of model of any system—be it static or dynamic—and if we are to understand the nature of operational research we must certainly, as rational persons, construct at least that.

In addition to the factor of *incidence* which creates a dichotomy between occasional and recurrent problems, we may apply the factor of *structure* at three levels: well-structured, part-structured, and ill-structured. The meanings of these terms should by now be reasonably clear. However, we may note that the more a problem can be structured, the more can the method of solution be reduced to a "rule-book" which can then either be applied by competent but relatively poorly-skilled persons or be made into a computer program. Well-structured problems are usually (but not invariably) *programable*, to use Professor Simon's term (Ref. 6). The beginnings of a classification are now to hand and an initial table can be constructed as follows:

Table 1. Tentative Classification of Problems

	Occasional	Recurrent
Well-structured	Eg: Queueing studies in city passenger terminal design (section 2)	Eg: Forecasting of passenger traffic for planning (section 5)
Part-structured	χ	Eg: Design of optimal advertising media schedules (section 4)
Ill-structured	χ	χ

I must comment upon the marginal classification of "occasional" problems. Such problems have traditionally been, and undoubtedly still are, usually solved by managers or their general staff assistants as part of their normal duties. This is appropriate wherever such problems are not well-structured. When, however, they are well-structured and, as in the case of the queueing problems that I have described, some special knowledge is called for to use a formula or an algorithm correctly and expeditiously, it is better that the calculations be assigned to people who have such knowledge than that amateurs attempt them. This then is a role which operational research practitioners may be asked to play. It is not research, of course, but it is a useful service and not very time-consuming. If many problems occur which are individually occasional with respect to subject matter, but are generically similar (e.g., straightforward applications of queueing formulas) and *collectively* numerous, then it is clear that computer programs may be justifiably written to solve them; and the main burden of administering their use may be transferred to a computer group. (This is our procedure in Air Canada.) This has such advantages both as regards speed and economy that, subject to economic justification, it is desirable for an operational research group to identify generic problems that are currently ill-structured and by research attempt both to structure them and develop general, programmed solutions to them. This is indeed what true "theoretical" operational research may be said to be about. Its involvement with computer science is obvious. Similar arguments apply with at least equal force to recurrent, ill-structured problems.

Recurrent, and even more so *routine*, problems are much less reasonable concerns of operational research practitioners once the research work has produced a structure or model, the development work has produced programs and procedures and monitored, pilot tests have proven the applicability of the solution procedures. If an O. R. group is not to change character and become a production rather than a research unit, such *systems* must be turned over to be manned and operated routinely by the line or staff service departments responsible for the subject matter which they relate to. This raises two issues which are significant in practice.

First of all, much of the detailed work of devising operable systems based upon models and techniques devised by operational research workers is of a type which many people of the latter type are ill-equipped in virtue both of training and temperament. My observations suggest that not all—and indeed, I am tempted to say, not many—people who are good at, and like, research and model development because of its content of conceptualization and analysis can be expected to enjoy writing procedures manuals. Nor can they be expected to acquire sufficient interest and expertise in ergonomics to design new work-positions and layouts which effective implementation of their proposals may call for. This is particularly needed when the subject matter concerns operational systems such as those discussed in the second section. Consequently, the involvement of "systems-and-procedures" analysts and industrial engineers in the progress of operational research studies from the development phase onwards is strongly indicated. In terms of organization such specialists may be drawn from separate specialist departments; included for such purposes in the operational research department itself (which may then consist of a "research" division and a "devlopment" division); or all such specialist management service groups, including operational research, should be collected together as one Management Advisory Services or Management Sciences department. The third course, which has been well argued by Brough, seems to be increasingly popular, in the airline industry at least. The first is probably the least desirable in most situations, as it easily generates intergroup rivalry and conflict and poses problems of the planning and coordination of effort. I have observed that the second can be as effective as the third in organizations where engineering and data-processing activities are organizationally dispersed or barely exist, but as ineffective as the first in other situations.

Second, the systems resulting from management science work frequently have a hard core of sophisticated thought and mathematics. The ultimate users in management or staff positions may reject them, or having accepted them later distort them, unless they themselves possess skills and knowledge of a fairly high order. What I am trying to say is that operational research can only be fully effective when corporate management is professional. In my opinion the growing importance of, and recognition granted to, University Schools of Business is a most encouraging phenomenon that promises well for the future of operational research. As more and more of their graduates enter, and work their way up the management ranks in, our business, industrial and governmental organizations it will become easier to carry out effective operational research within them. It is, I believe, still significant that operational research originated in military organizations which have long been essentially professional in character.

These have been necessary and, I trust, not irrelevant digressions, but it is time to return to our main theme: the types of operational research. I have briefly classified operational research in terms of its purposes and uses. More elaborate classifications can be built upon this foundation. For example, a three-way classification of problems in the field of management systems may be introduced: those related to *planning*, those related to *operational* management and those related to *monitoring* or control. However, I have convinced myself that from the viewpoint of operational research approaches these distinctions are primarily significant only for the design and physical implementation of *management information systems*. This is too large a topic to be dealt with here and so, with considerable regret, I must put it aside for discussion on some other occasion.

It is better to conclude by taking another view of operational research, to attempt to picture it in the round, by considering the levels of concept and models that it employs. Operational research, as I have stated more than once, has for its subject matter organized, man-made systems of men, machines, materials and money possessing strucure, content, communication and control. Such systems are observably evolutionary, dynamic and—sometimes at least—adaptive. A. J. Toynbee, in his massive work *A Study of History*, has provided an enormous quantity of evidence to show that such has been true of the members of one class of organizations, namely those which he defines to be *civilizations*. Part of the trouble which people appear to have in understanding his work— and I think this includes many professional historians—is that they bring to their reading of it conceptual models of one level, whereas,

Toynbee has attempted to analyze the course of history in terms of models of a different, higher level. It is the same thing on a humbler scale with the understanding which many people, even operational research practitioners themselves, have brought to bear upon the work of cyberneticians.

Models may be classified conveniently in two ways. The first is whether they are deterministic or probabilistic. These terms need no explanation. However, it is important to note that each alternative may be joined to each of the classes listed below which relate to the second way of classification.

1. *Frameworks*, which are static models and schemes of classification and ordering. Classical organization theory deals principally with such models.

2. *Clockworks*, which are dynamic models taking into account the passage of time. Newton's universe (Book III of the *Principia*) is a deterministic clockwork.

3. *Thermostats*, or regulated clockworks, which provide for the restoration of an unique equilibrium preset by an external agency in the event of environmental perturbations.

4. *Adaptive Models*, which provide for shifts to alternative equilibria (possibly drawn from a continuous range) in response to environmental changes in order to optimize the performance of some function.

Models of greater complexity than these are easily conceivable, but these four will suffice for the present.

The majority of concepts in use by managements and staff specialists in organizations at the present time are probably of the first two classes and probably deterministic: we may say they are of types 1-D and 2-D. It is my personal experience that few of the older generation of businessmen and government officials have much ability to think in terms of the conceptual levels 1-P and 2-P, using P to denote "probabilistic." The majority of management control systems which I have observed in use appear to be based upon a view of organizations as deterministic thermostats, level 3-D. Accountants have traditionally set up expense reporting procedures along these lines, though in recent years they have been increasingly cognizant of the importance of random events and have introduced concepts of the 3-P type into use, above all perhaps in auditing. Industrial quality control methods have, of course, been based increasingly upon type 3-P concepts in the past two decades. On the whole, however, there is little evidence of the use of either deterministic or probabilistic adaptive models. This may seem strange, as observation of the behavior of organizations suggests very strongly that they are, individually, consciously striving to be adaptive. The fact of the matter is, however, that we are in a relatively poor position theoretically to design probabilistic adaptive models, as a reading of some of the books quoted in the bibliography will confirm.

Operational research has concentrated upon the use of models of the types 1-D, 1-P, 2-D, and 2-P itself for the most part of its existence, and when a model of some higher level type has been indicated, various devices have been used to reduce it to one of these four. The case-histories which I have described to you illustrate this proposition. For,

1. The queueing processes studied as described in the second section were in reality of at least the 2-P type, but by application of the device of "statistical equilibrium," they were reduced to models of the 1-P type. That is stochastic processes were reduced to problems in statistical distribution theory: time-effects were suppressed. In practice, as I have pointed out in my book *Applied Queueing Theory*, the check-in process at least was of the type 3-P, as the human operators used their observation of queue-lengths as feedback control upon their rate of work.

2. The advertising example which I discussed in the fourth section was studied by a model at the 1-D level: neither time nor statistical effects were considered. (In later work, a deterministic dynamic model of the 2-D level was constructed to solve a more intricate problem.)

3. The *EMMA* model of the fifth section in itself is static but prob-abilistic, at the 1-P level. However, the *system* in which it is imbedded, which is of the part-structured, recurrent variety, is conceptually of the 2-P but it does function in time.

4. The final case which I described relates to a forecasting system with particularly interesting features. The real process is clearly of the high 4-P level. However, to model this would have required a most sophisticated use of the historical record. By the device of exponential smoothing, an essentially non-Markov process was made to look like a Markov process in discrete time, as all historical data is condensed into the current state-variable, and the next forecast value of this state-variable depends only upon its present value. Furthermore the use of SQC techniques in relating the actual to the predicted value of the state-variable, and hence permitting feedback control by the human users, introduced adaptivity. The resulting *system* is decidedly based upon concepts of the 3-P type.

Overall, then, a tentative framework for operational reseach may be built up by combining the classification by problem type with that by level of model concept. This is not a task for which we have time now, although the procedure is apparent enough.

In this paper I have attempted to portray the many-sided activity of operational research by means of "snapshots" taken from different viewpoints at different angles, and with different exposures. I hope that this discontinuous form of presentation has served its purpose and that the implied omissions, incomplete expositions, and undeveloped arguments will stimulate the audience to read more, think more, and above all to do more about operational research.

ACKNOWLEDGMENT

The case history material of section 2 is based upon work done in BEA in association with Dr. P. A. Longton (now professor of marketing at the University of Western Australia) and others. Section 4 derives from research carried out—again in BEA—by John Burkart and myself. Section 5 is based upon work done in Air Canada by Dr. I. Elce, Dr. K. Haas, R. W. Linder, H. J. G. Whitton, and, to a lesser extent, myself. To all of these associates I express my indebtedness. Operational research is always a team-effort and I always have been fortunate in my colleagues.

One of the most stimulating and interesting results of this presentation was the opportunity afforded to talk with people who attended them and who arranged numerous social functions where such discussions could take place. The list of their names would be very long, too long to give here, but to all of them my wife and I are very grateful. In particular, however, I must thank Dr. Ramakrishnan and his charming wife, Prof. Srinivasan and Dr. Ranganathan of the Indian Institute of Technology, and Mr. C. Subramanian, chairman of the government of India Aeronautics Committee and president of the Tamil Nad Congress Committee, for introducing me to the problems of planning and development in his country. In addition we were stimulated, instructed, and entertained by the distinguished Visiting Fellows from the United States: Professor Roland Good (Iowa), Professor Gordon Shaw (California), Dr. L. Rubel (Institute for Advanced Study, Princeton), and Professor Fuchs (Cornell). To them also I express my appreciation.

REFERENCES*

1. R. Ackoff and P.B. H. Rivett, *A Manager's Guide to Operational Reseach*, John Wiley and Sons, New York, 1965.
2. W. Ross Ashby, *An Introduction to Cybernetics* Chapman and Hall, London, 1956.
3. R. Ackoff (Ed), *Progress in Operations Research, Vol. I*, John Wiley and Sons, New York, 1959.
4. D. B. Hertz and R. T. Eddison (Eds), *Progress in Operations Research, Vol. II John Wiley and Sons*, New york, 1964.
5. A. M. Lee *Applied Queueing Theory*, Macmillan and Co., London, 1966.
6. H. A. Simon, *The New Science of Management Decision*, Harper and Row, New York, 1960.
7. A. M. Lee, "Decision rules for media scheduling: static campaigns" *Oper Res Quart*, **13** (3): 229–241 (1962).
8. A. M. Lee, "Decision rules for media scheduling: dynamic campaigns" *Oper Res Quat*, **14** (4): 365–372 (1963).
9. D. M. Ellis, "Building up a sequence of optimum media schedules" *Oper Res Quart*, **17** (4): 413–424 (1966).
10. I. Elce, A. M. Lee, and R. W. Linder, "Forecasting models for corporate planning" *paper presented at 1967 Conf of Can Oper Res Soc; to appear in J. Can. O. R. Soc.*
11. H. J. G. Whitton and R. W. Linder, "Computer forecasting of passenger flows" *Proc. Second AGIFORS Symposium*, American Airlines, New York, 1962.

---

*The literature of operational research is already very extensive and the works listed below constitute only a minute selection. Mainly, they are either books suitable for readers new to the field or books and articles having some direct bearing upon the content of this paper. The omission of any title implies no criticism of its value.

Nevanlinna Theory and Gap Series

W. H. J. FUCHS

CORNELL UNIVERSITY*
Ithaca, New York

---

1. INTRODUCTION

A very attractive feature of complex variable theory is the interplay of the approach via power series (Weierstrass) and the potential theoretical approach (Riemann). The Nevanlinna theory studies the distribution of values of meromorphic functions, by potential–theoretic methods. In this paper we shall make an application of this theory to power series.

2. RECALL OF SOME DEFINITIONS

For any meromorphic function $g(z)$ we define

$$m(r, g) = \frac{1}{2\pi} \int_0^{2\pi} \log^+ |g(re^{i\theta})| d\theta \qquad (r \geq 0)$$

where $\log^+ u = \max(\log u, 0)$. If $n(t, c)$ is the number of roots of $g(z) = c$ in $|z| \leq t$ (each zero counted with its proper multiplicity), then

$$N(r, c; g) = \int_0^r \frac{n(t, c) - n(0, c)}{t} dt + n(0, c) \log r$$

If $f(z)$ is an entire function, one can prove that for all complex numbers c with the possible exception of those in a small exceptional

---

*Department of Mathematics.

set

$$N(r, c; f) \sim m(r, f)$$

A value c is called a *deficient value* with respect to the entire function $f(z)$, if

$$\lim_{r \to \infty} \sup N(r, c; f)/m(r, f) = 1 - \delta(c, f) < 1$$

An alternative way of writing the "deficiency" $\delta(c, f)$ is

$$\delta(c, f) = \lim_{r \to \infty} \inf m \left(r, \frac{1}{f - c}\right)/m(r, f) \tag{1}$$

The function $m(r, f)$ is called the characteristic function of the entire function $f(z)$. The characteristic function is an increasing function of r and if $f(z)$ is not a polynomial

$$\log r = o(m(r, f)) \qquad (r \to \infty)$$

A set of functions $\{\phi_k(t)\}_{k=1}^{\infty}$, where $\phi_k(t)$ belongs to the space $L^2(a, b)$ of complex valued, square-integrable functions defined on $a < t < b$ is *complete* in $L^2(a, b)$, if the relations

$$h(t) \in L^2(a, b), \int_a^b h(t)\phi_k(t)dt = 0 \qquad (k = 1, 2, \ldots)$$

imply $h(t) = 0$ almost everywhere in (a, b).

3. STATEMENT AND PRELIMINARY DISCUSSION OF THEOREM

We shall prove the following:

Theorem: *Let $0 < n_1 < n_2 \cdots$ be a sequence of positive integers such that*

$$\{e^{i n_k t}\}_{k=1}^{\infty}$$

is not complete in $L^2(-D, D)$. Let

$$f(z) = c_0 + c_1 z^{n_1} + c_2 z^{n_2} \cdots \tag{2}$$

be an entire function of finite lower order μ.
There is an absolute constant K and a constant

$$A(\mu) < K(1 + \mu \log^+ \mu) \tag{3}$$

such that for every complex number w

$$\delta(w, f) < A(\mu)D \tag{4}$$

Discussion: By a well-known theorem of Paley–Wiener, a function $H(z)$ can be written in the form

$$H(z) = \int_{-D}^{D} h(t)e^{izt}\,dt \qquad [h \in L^2(-D, D)]$$

if and only if $H(z)$ is an entire function of the complex variable z satisfying

$$|H(z)| < Ae^{D|z|} \qquad H(x) \in L^2(-\infty, \infty) \tag{5}$$

The hypothesis that $\{e^{in_k t}\}$ is not complete in $L^2(-D, D)$ can therefore be rephrased: There is an entire function $H(z)$ satisying (5) such that

$$H(n_k) = 0 \qquad (k = 1, 2, \dots) \tag{6}$$

Without loss of generality we may also suppose that

$$H(0) \neq 0 \tag{7}$$

[otherwise replace $H(z)$ by $H(z)z^{-p}$].
  If

$$\sum_{k=1}^{\infty} \frac{1}{n_k} < \infty \tag{8}$$

then

$$H(z) = \prod_{j=1}^{M} (z - n_j) \prod_{k>M} \frac{\sin(\pi z/n_k)}{(\pi z/n_k)}$$

satisfies (5) with

$$D = D_M = \pi \prod_{k>M} \frac{1}{n_k} + \epsilon$$

for every $\epsilon > 0$. By choosing M large, D_M can be made arbitrarily small. We obtain therefore from theorem 1 the following:

Corollary: If the entire function (2) *has finite lower order and if* (8) *holds, then the function* (2) *has no finite deficient values.*

This is a partial sharpening of a theorem due to Biernacki: an entire function of the form (2) for which (8) holds takes every value infinitely often.
  If q is a positive integer and $n_k = kq(k = 1, 2, \dots)$, then

$$H(z) = (q/\pi z)\sin(\pi z/q)$$

satisfies (5) with $D = \pi/q$. By applying theorem 1 to $f(z) = \exp(z^q)$

we find

$$1 = \delta(0, f) < A(q)(\pi/q)$$

This yields a lower bound for $A(\mu)$ when μ is an integer. It seems likely that $A(\mu) = O(\mu)$.

For functions whose order is an integer, F. Sunyeri Balanguer [*Acta Math.* **87**: (1952)] has obtained results which are of a nature similar to theorem 1.

Proof of Theorem 1: There is nothing to prove if $f(z)$ reduces to a polynomial. We may therefore assume that $\log r = o(m(r, f))$. It is enough to prove the theorem for $w = 0$, the general case then follows by considering $f(z) - w$. We may further suppose that $f(0)$ is equal to an assigned positive number C. If this is not the case to start with, replace $f(z)$ by $Bf(z) z^{-p}$, where z^p is the lowest power with a nonvanishing coefficiet in the expansion (2). This replacement does not essentially change the zeros or the magnitude of the Nevanlinna characteristic. Also, the set $\{e^{i(n_k - p)t}\}$ is not complete in $(-D, D)$, if $\{e^{i n_k t}\}$ is not complete [replace $h(t)$ by $h(t)e^{ipt}$ in the definition of completeness].

By our hypothesis it is possible to find an entire function $H(z)$ satisfying (5), (6), and (7). The function $H(z)$ has a representation

$$H(z) = \int_{-D}^{D} h(\theta) \, e^{iz\theta} \, d\theta \qquad h(\theta) \in L^2(-D, D)$$

Let γ be real and consider

$$E = \int_{-D}^{D} f(re^{i(\gamma+\theta)}) \, h(\theta) d\theta$$

$$= c_0 \int_{-D}^{D} h(\theta) d\theta + c_1 r^{n_1} e^{i\gamma n_1} \int_{-D}^{D} h(\theta) e^{i n_1 \theta} d\theta + \cdots$$

$$= c_0 H(0) + c_1 r^{n_1} e^{i} r^{n_1} H(n_1) + \cdots$$

$$= c_0 H(0)$$

in view of equation (6). Also,

$$|E| \leq \sup_{\gamma - D \leq \phi < \gamma + D} |f(re^{i\phi})| \int_{-D}^{D} |h(\theta)| d\theta$$

The integral on the right-hand side is finite, by Schwarz's inequality. Hence,

$$\sup_{\gamma - D \leq \phi \leq \gamma + D} |f(re^{i\phi})| \geq |E| \Big/ \int_{-D}^{D} |h(\theta)| d\theta = |c_0 H(0)| \Big/ \int_{-D}^{D} |h(\theta)| d\theta$$

By an earlier remark we may suppose that $|c_0| = |f(0)|$ is so large that the right-hand side of the last inequality is equal to 1. In other words we have shown that $|f(z)| \geq 1$ at some point of every arc of $|z| = r$ which subtends an angle $\geq 2D$ at the origin.

Now

$$m(r, 1/f) = -\frac{1}{2\pi} \int_F \log f|(re^{i\theta})| \, d\theta$$

where $F = F(r)$ is the set of θ (mod 2π) for which $\log |f(re^{i\theta})| \leq 0$. $F(r)$ is the union of a finite number of intervals (α, β) at whose end-points $\log |f(re^{i\theta})| = 0$. If r is chosen so that there are no zeros of $f(z)$ on $|z| = r$, then by integration by parts in any typical interval (α, β),

$$-\frac{1}{2\pi} \int_\alpha^\beta \log |f(re^{i\theta})| \, d\theta = \frac{1}{2\pi} \int_\alpha^\beta \left(\theta - \frac{\alpha + \beta}{2} \right) \frac{d}{d\theta} \log |f(re^{i\theta})| \, d\theta$$

(9)

But

$$\log |f(re^{i\theta})| = Re \, \log f(re^{i\theta})$$

and so

$$\left| \frac{d}{d\theta} \log f(re^{i\theta}) \right| = \left| Re \, \frac{d}{d\theta} \log f(re^{i\theta}) \right|$$

$$\leq \left| \frac{d}{d\theta} \log f(re^{i\theta}) \right| = r \left| \frac{f'(re^{i\theta})}{f(re^{i\theta})} \right| \quad (10)$$

We also know that the length of the interval (α, β) can not exceed $2D$. Therefore, by equations (9) and (10)

$$-\frac{1}{2\pi} \int_\alpha^\beta \log |f(re^{i\theta})| \, d\theta < \frac{Dr}{2\pi} \int_\alpha^\beta \left| \frac{f'(re^{i\theta})}{f(re^{i\theta})} \right| d\theta$$

$$m(r, 1/f) = -\frac{1}{2\pi} \Sigma \int_\alpha^\beta \log |f(re^{i\theta})| \, d\theta < \frac{Dr}{2\pi} \int_0^{2\pi} \left| \frac{f'(re^{i\theta})}{f(re^{i\theta})} \right| d\theta \quad (11)$$

By a known result* there are arbitrarily large values of r for which

$$r \int_0^{2\pi} \left| \frac{f'(re^{i\theta})}{f(re^{i\theta})} \right| d\theta < K(1 + \mu \log^+ \mu) \, m(r, f)$$

The theorem now follows from (11) and from the definition (1) of deficiency.

---

*W. H. J. Fuchs, *Ann. Math.* **68** (2): 203–209 (1958).

Uniform Distribution and Densities on Locally Compact Abelian Groups

L. A. RUBEL

UNIVERSITY OF ILLINOIS   and   INSTITUTE FOR ADVANCED
Urbana, Illinois   STUDY
                                                  Princeton, New Jersey

In this paper we shall discuss the uniform distribution of a sequence on a locally compact Abelian group and the notion of a density on the same kind of group, and describe a number of results about them that have recently been obtained, and the relationship between the two.

In 1916, Weyl introduced the notion of a sequence $\{x_v\}$ of real numbers being uniformly distributed mod 1 as meaning that if $((x_v))$ denotes the fractional part of x_v, then for each closed subinterval I of the unit interval $(0, 1)$, $\lim_{n \to \infty} I(n)/n = |I|$, where $I(n)$ counts the number of elements of the sequence $\{((x_v))\}$ that lie in I, for $v = 1, 2, \ldots,$ n, and $|I|$ denotes the length of I. He established the so-called Weyl criterion that $\{x_v\}$ is uniformly distributed (u.d.) if and only if

$$\lim_{n \to \infty} \frac{1}{n} \sum_{v=1}^{n} e^{2\pi i x_v k} = 0$$

for each nonzero integer k.

In 1943, Eckmann introduced the notion of uniform distribution on a compact group G, of which uniform distribution mod 1 is the special case corresponding to the circle group T in the usual topology. A sequence $\{x_v\}$ of elements of G is said to be u.d. if $\lim_{n \to \infty} I(n)/n = |I|$ for each closed set I whose boundary has Haar measure zero, where $I(n)$ counts the number of x_ν that belong to I for $\nu = 1, 2, \ldots, n$ and $|I|$ denotes the Haar measure of I. He proved the analogue of the Weyl criterion, that $\{x_\nu\}$ is u.d. if and only if

$$\lim_{n \to \infty} \frac{1}{n} \sum_{\nu=1}^{n} \chi(x_\nu) = 0 \tag{1}$$

for each nontrivial continuous character χ on G. To say that $\{x_\nu\}$ is u.d. on a compact group G is to say that

$$\frac{\epsilon_{x_1} + \epsilon_{x_2} + \cdots + \epsilon_{x_n}}{n} \xrightarrow{w} \lambda, \tag{2}$$

that is, that the Césaro means of the unit point masses located at the x_ν tend weakly to Haar measure λ. Eckmann also introduced and characterized the so-called monothetic groups which are those groups that have a dense sequence of the form $\{nx\}$, $n = 0, \pm 1, \pm 2, \ldots$.

There was some interest in finding the correct definition of uniform distribution on locally compact Abelian groups that are not necessarily compact. In 1961, Niven introduced the notion of uniform distribution on the group $\mathscr{Z}$ of integers by saying, roughly, that a sequence $\{x_\nu\}$ of inegers is uniformly distributed if the $\{x_\nu\}$ fall into each arithmetic progression I with limiting frequency equal to the arithmetic density of I, that is, if $\lim_{n \to \infty} I(n)/n = |I|$, where $I(n)$ is as before and $|I| = 1/a$ if $I = \{ax + b : x = 0, \pm 1, \pm 2, \ldots\}$. Several mathematicians proposed the definition for $\mathbb{R}$, the group of real numbers in the usual topology, that $\{x_\nu\}$ be u.d. if $\{x_\nu\}$ is u.d. mod t for every nonzero real number t.

In 1965, Rubel [L. A. Rubel, "Uniform distribution in locally compact Abelian groups," *Comm. Math. Helv.* **93**: 253–258 (1965)] published a definition of u.d. on a locally compact Abelian (L.C.A.) group, that appears now to be correct. In case G is compact, or $G = \mathscr{Z}$, or $G = \mathbb{R}$, it coincides with the previous definitions. We say that $\{x_\nu\}$ is u.d. in G if $\{\phi(x_\nu)\}$ is u.d. in Q for each compact quotient Q, where $\phi: G \to Q$ is the canonical homomorphism. The Weyl criterion (1) holds, where we now require that the characters χ be periodic, a notion to be discussed in a moment. But what is to take the place of (2)? Surely (2) cannot carry over directly, since Haar measure on a noncompact group has infinite variation. The answer to this is to replace Haar *measure* by Haar *density*. This leads us to introduce and study the notion of a density on an L.C.A group.

In what follows, we will suppose that G is an L.C.A. group. We say that a closed subgroup H of G is of compact index in G to mean that G/H is compact. A function f on G is periodic if it is constant on the cosets of some subgroup H of compact index, and we then say that f has period H or sometimes period G/H. A continuous periodic function f is characterized by having a compact orbit, where $\text{Orb} f = \{f_t : t \in G\}$ where $f_t(x) = f(x + t)$, and the topology is the uniform topology. We recall that a function is almost periodic if its orbit has compact closure. We remark that on $\mathbb{R}$ the periodic functions are those that are periodic in the usual sense, but on $\mathbb{R}^n$, the periodic functions in our sense are those that are n-fold periodic in the usual sense. A character χ is periodic if and only if $G/\ker \chi$ is compact. This happens exactly when χ is a discrete element of the dual group $G^\wedge$, that is, the group generated by χ is a discrete subgroup of $G^\wedge$.

The first question is: which groups G contain u.d. sequences? This was answered by I.D. Berg, M. Rajagopalan, and Rubel in *Trans. Amer. Math. Soc.* **133**: 435–446 (1968).

Definition: The L.C.A. group G is K-separable if and only if there exists a sequence $\{g_v\}$ of elements of G such that for each compact quotient Q of G, $\{\phi_Q(g_v)\}$ is dense in Q, where $\phi_Q: G \to Q$ is the canonical homomorphism.

Theorem: There exists a u.d. sequence in G if and only if G is K-separable.

This is not an entirely satisfactory result. They further study this question more extensively using, in particular, the structure theorem for L.C.A groups, that there exists an open and closed subgroup H of G such that $H = K \times \mathbb{R}^n$ where K is compact and $\mathbb{R}^n$ is Euclidean n-space. Let us write Γ^p for the set of periodic continuous characters, and $\Gamma = G^\wedge$ for the group of all continuous characters.

Lemma: $\Gamma^p = \Gamma$ if and only if each discrete quotient of G is of bounded order.

Theorem: Γ^p forms a group if and only if either every discrete quotient of G is of bounded order or else G is totally disconnected.

Theorem: If card $\Gamma^p \leq C$ then there exists a u.d. sequence in G. If Γ^p is a group and there exists a u.d. sequence in G, then card $\Gamma^p \leq C$. For any cardinal number N, there is an L.C.A. group G that admits a u.d. sequence and yet card $\Gamma^p \geq N$.

Here, card X denotes the cardinal number of X and C denotes the cardinal number of the continuum. The last theorem is based on the assumption of the continuum hypothesis.

The second question answered in the above paper is when uniform distribution on the L.C.A group G is just equivalent to uniform distribution on some one associated compact group.

Theorem: *Let G be an L.C.A group that supports a u.d. sequence and suppose that Γ^p is a group. Then $\{g_v\}$ is u.d. in G if and only if $\{\phi(g_v)\}$ is u.d. in $\bar{G}^p$, where ϕ is the canonical map of G into its semiperiodic compactfication $\bar{G}^p$ that is the dual of the discrete group generated by the periodic continuous characters on G. Also, $\bar{G}^p$ is the only compactification of G with this property. If Γ^p is not a group, then no compactification will serve.*

Definition: *The L.C.A group G is monogenic if there exists an $x \in G$ such that $\{nx\}, n = 1, 2, 3, \ldots$ is u.d. in G.*

The next result was proved by M. Rajagopalan and J.J. Rotman in "Monogenic groups," *Compositio Math.* **18**: 155–161 (1967).

Theorem: *A discrete group G is monogenic if and only if $G = D \oplus H$ where D is divisible and H is a pure subgroup of a universal monogenic group that contains a main diagonal.*

This is based on the following definition:

Definition: *Let P a set of distinct primes, and for each $p \in P$ let C_p denote either a p-primary cyclic group or the additive group of p-adic integers. We call $\prod C_p$ a universal monogenic group. An element $x \in \prod C_p$ is called a main diagonal if, for all p, its p-th coordinate is a generator of C_p when C_p is cyclic, or a p-adic unit otherwise.*

Recently, Rajagopalan (unpublished) has characterized the structure of the general L.C.A. monogenic group. His description is complicated, and we do not give it here.

We now turn to the study of densities on L.C.A groups. A density is a compatible bounded system of measures on the compact quotients of G. The notion of density is studied in a paper by Berg and Rubel, "Densities on locally compact Abelian groups" to appear in *Ann. Inst. Fourier* (Grenoble).*

---

*See research announcement in: *Bull. Amer. Math. Soc.* **74**: 298–300 (1968).

Definition: A density μ on G is a system $\{\mu_Q\}$ of measures μ_Q on the compact quotients Q of G that satisfy the following conditions:

1. $\|\mu\| = \sup \{\|\mu_Q\|\} < \infty$
2. $\mu_Q(E) = \mu_{Q'}(\phi^{-1}(E))$ *for any Borel set E in Q*

where $\phi: Q' \to Q$ is the canonical homomorphism whenever Q is a quotient of Q'.

The densities form a Banach algebra $D(G)$, that coincides with the measure algebra $M(G)$ when G is compact and not otherwise. The convolution $\mu*\nu$ is defined by $(\mu*\nu)_Q = \mu_Q*\nu_Q$. The simplest density is Haar density λ defined by $(\lambda)_Q = \lambda_Q$, where λ_Q is Haar measure on Q. Haar density on $\mathscr{Z}$ coincides with the natural density that assigns to $\{ax + b\}$ the density $1/a$. The algebra $D(G)$ also has the weak topology with respect to the continuous periodic functions, where

$$\int f\,d\mu = \int f_Q\,d\mu_Q$$

where Q is of period f, and $f_Q(x + H) = f(x)$ if $Q = G/H$. We can now express the form equivalent to (2) for L.C.A. groups.

Theorem: The sequence $\{x_\nu\}$ in G is u.d. if and only if

$$\frac{\epsilon_{x_1} + \epsilon_{x_2} + \cdots + \epsilon_{x_n}}{n} \xrightarrow{w} \lambda$$

where λ is Haar density on G and ϵ_{x_i} is the unit point density located at x_i.

It is not hard to prove that if $f = f_1 + f_2 + \cdots + f_n$ is a sum of n periodic functions and μ is a density, then $\int f\,d\mu$ is well defined by

$$\int f\,d\mu = \int f_1\,d\mu + \int f_2\,d\mu + \cdots + \int f_n\,d\mu.$$

A continuous function f on G is semiperiodic if it lies in the uniformly closed span of the periodic functions, or equivalently, if it lifts to a continuous function on $\bar{G}^p$. A basic question is, when do all densities extend to bounded linear functionals on the semiperiodic functions? This can be shown to be equivalent to each density corresponding in the natural way to a measure on $\bar{G}^p$.

Theorem: Each density extends to a bounded linear functional on the semiperiodic functions precisely when the sum of each two periodic continuous characters on G is periodic.

We can construct a group G on which the sums of periodic continuous characters are periodic but for which there exist two periodic continuous functions with a nonperiodic sum.

Theorem: The sum of each two periodic continuous characters on G is periodic precisely when either G is totally disconnected or G has no $\mathbb{R}^n$ part and every discrete quotient of G has bounded order.

Definition: Two subgroups H_1, H_2 of compact index in G are said to be incident if there is a proper subgroup H of compact index such that $H \supseteq H_1 \cup H_2$. Two subgroups are said to be independent if there is no third subgroup that is incident to both of them.

Theorem: The sum of each two periodic continuous characters on G is periodic precisely when G contains no two independent subgroups of compact index.

In the sequel, we use the theory of almost periodic functions and their Fourier expansions.

Lemma: If $f \in$ A.P. and $f \sim \sum a_i \chi_i$, $a_i \neq 0$, then f is periodic if and only if the group $[\{\chi_i\}]$ generated by the $\{\chi_i\}$ is a discrete subgroup of $G^\wedge$.

Theorem: Suppose the sum of any two periodic continuous characters on G is periodic. Then any finite linear combination is.

Proposition: If $f_1, f_2, \ldots, f_n$ are continuous periodic continuous functions, then for any density μ

$$\left| \sum_{i=1}^{n} \mu(f_i) \right| \leq 2^{n+1} \|\mu\| \left\| \sum_{i=1}^{n} f_i \right\|_\infty$$

Proposition: On $\mathbb{R}$, each real density decomposes as the difference of two positive densities.

Proposition: On the cyclinder group $\mathbb{R} \times T$, where T is the circle group, there is a real density that has no decomposition.

We denote the complex numbers by **C**.

Definition: The Fourier transform $\mu^\wedge$ of the density μ is that function $\mu^\wedge : \pi \to \mathbf{C}$, where π is the set of periodic continuous characters on G, defined by

$$\mu^\wedge(\chi) = \int \chi \, d\mu.$$

Definition: *A function* $\phi: \pi \rightarrow \mathbf{C}$ *is called quasi-positive definite if*

$$\sum_{i,j=1}^{n} \phi(\chi_i \bar{\chi}_j) \xi_i \bar{\xi}_j \geq 0$$

where $\{\chi_i\}$ *is any set of periodic continuous characters on* G *that have a common period.*

We now have the analogue of Bochner's theorem.

Theorem: *The function* $\phi: \pi \rightarrow \mathbf{C}$ *is quasi-positive-definite if and only if* ϕ *is the Fourier transform of a positive density.*

We now think of G as having the relative topology induced by its map into $\bar{G}^p$. If translation is a norm continuous operation on the density μ, then we say that μ is absolutely continuous and we denote by $D^1(G)$ the Banach algebra of all such densities. If G is compact, then $D^1(G)$ coincides with $L^1(G)$, but not otherwise. We suppose that the product of periodic characters is periodic.

Theorem: *If* $\mu \in D^1(G)$ *then* $\mu^\vee(\chi) \rightarrow 0$ *as* χ *leaves finite subsets of* π.

This is the analogue for $D^1(G)$ of the Riemann–Lebesgue lemma.

Theorem: *The spectrum of* $D^1(G)$ *is* π. *That is, each complex homomorphism of* $D^1(G)$ *must be evaluation of the Fourier transform at some periodic continuous character* χ.

Theorem: *Spectral synthesis holds for* $D^1(G)$. *That is, each closed ideal in* $D^1(G)$ *is characterized by the common zero set of the Fourier transforms of its densities.*

Many questions remain open. For example, when does a given density extend, and when does it decompose? It is hoped that research will continue to investigate these problems.

Lectures on Vector Spaces of Analytic Functions

L. A. RUBEL

UNIVERSITY OF ILLINOIS and *INSTITUTE FOR ADVANCED*
Urbana, Illinois, *STUDY*
Princeton, New Jersey

INTRODUCTION

These lectures deal with the field of interaction between the theory of analytic functions and functional analysis. In the first section, we shall study the classical theory of analytic functions of a single complex variable, using functional analysis as a tool to obtain some basic classical theorems like Cauchy's integral theorem and the Mittag–Leffler theorem. In the second section, we dualize a gap theorem of C. Renyi for periodic entire functions to obtain some approximation theorems. One of these is a theorem about weighted polynomial approximation on the integers. In the third section we make a new application of perturbation of a basis in a Banach space to prove uniqueness theorems for analytic functions of one and of several complex variables. In the fourth section, we sketch a theory of duality of vector spaces of entire functions that leads to the solution of the problem of spectral synthesis for mean-periodic entire functions of one complex variable. The remainder of the lectures is devoted to a study of the space of bounded analytic functions in various weak topologies.

1. THE SPACE $H(G)$

Let G be an open subset of the complex plane $\mathscr{C}$, and denote by $H(G)$ the space of all holomorphic (i.e., analytic) functions on G. We put on $H(G)$ the topology of uniform convergence on compact subsets of G. A neighborhood of $f \in H(G)$ has the form

$$N = N(f; K, \epsilon) = \{g \in H(G) : |f(z) - g(z)| < \epsilon \text{ for all } z \in K\},$$

where K is a compact subset of G and ϵ is an arbitrary positive number. It is clear that $H(G)$ is a locally convex topological algebra. The topology is also specified by the family of seminorms

$$\|f\|_K = \sup \{|f(z)| : z \in K\}.$$

Since there exists an exhausting sequence $\{K_n\}$ of compact subsets of G, that is, an increasing sequence $\{K_n\}$ such that each compact subset of G is contained in some K_n, $H(G)$ is consequently a metric space with the metric

$$\rho(f, g) = \sum 2^{-n} \frac{\|f - g\|_{K_n}}{1 + \|f - g\|_{K_n}}.$$

Equivalent to Montel's theorem on normal families is the result that the compact subsets of $H(G)$ are precisely the closed and bounded sets. Recall that a set in a locally convex topological vector space is bounded when some homothety of each neighborhood of 0 covers it. A bounded set in $H(G)$ is just a set of functions that is uniformly bounded on each compact subset of G.

Now $H(G)$ is a closed subspace of the space $C(G)$ of all continuous complex valued functions on G in the same topology. By the Riesz Representation theorem, the dual of $C(G)$ is the space of all bounded Borel measures of compact support in G. By the Hahn–Banach theorem, then, the dual $H(G)'$ of $H(G)$ is the quotient space consisting of all equivalence classes of such measures where $\mu \sim \nu$ means that $\int f \, d\mu = \int f \, d\nu$ for all $f \in H(G)$. Given a continuous linear functional L on $H(G)$, we see that L has at least one extension to $C(K)$, where K is a compact set supporting one of the representing measures μ, and $C(K)$ is the space of continuous functions on K in the supremum norm topology. We still denote such an extension by L. We desire a better representation of $H(G)'$. For this, we require the following weak form of the Cauchy integral theorem and Cauchy integral formula. (See Saks and Zygmund, *Analytic Functions*, Warsaw, 1965, Second Edition, p. 155.)

1.1 Theorem: Let G be an arbitrary open set in $\mathscr{C}$ and let K be a compact subset of G. Then there exists in $G \backslash K$ a finite union (cycle) Γ of simple closed curves, made up of finitely many oriented line segments *parallel to the coordinate axes, such that for each* $f \in H(G)$

(i)
$$\int_\Gamma f(z)\, dz = 0$$

(ii)
$$f(w) = \frac{1}{2\pi i} \int_\Gamma \frac{f(z)}{z-w}\, dz \qquad \text{for all } w \in K.$$

Further, if F is holomorphic on the complement of K on the Riemann sphere and $F(\infty) = 0$, then

(iii)
$$F(w) = -\frac{1}{2\pi i} \int_\Gamma \frac{F(z)}{z-w}\, dz$$

for w in the unbounded component of the complement of Γ.

We call this "the weak Cauchy theorem." The proof consists of throwing down a fine square mesh on the plane, choosing those squares that touch K, and then selecting those boundary segments of these touching squares, that lie on exactly one square of the system, and applying the Cauchy integral theorem and Cauchy integral formula for a rectangle.

By G' we denote the complement of G on the Riemann sphere, so that $\infty \in G'$. If A is any subset of the Riemann sphere, then a complex-valued function f which is holomorphic on an open set $B \supseteq A$ is said to be locally holomorphic on A. Two such functions f and g are said to be equivalent, written $f \sim g$, if there is an open set $B \supseteq A$ such that $f|_B = g|_B$. We denote by $[f]$ the class of functions equivalent to f, and call these equivalence classes germs of holomorphic functions on A. We denote by $H_0(G')$ the space (under the obvious pointwise operations) of germs of functions holomorphic on G' that vanish at ∞.

1.2 Theorem: *The dual space $H(G)'$ of $H(G)$ may be identified with $H_0(G')$, where the bilinear form pairing $H(G)$ and $H_0(G')$ is given by*

$$\langle f, [F] \rangle = \frac{1}{2\pi i} \int_\Gamma f(w)\, F(w)\, dw$$

for $f \in H(G)$ and $[F] \in H_0(G')$, where Γ is a cycle contained in the intersection of G with the domain of analyticity of the representative function F of $[F]$, that has winding number $+1$ around every singular point of F, and each component of which winds once around some singular point of F. If L is a continuous linear functional on $H(G)$, then the corresponding function F is given by $F(w) = L(1/w - z)$ for w in an *open set containing G' [where L denotes the functional extended to $C(G)$].*

For certain computations, it is easier to use the fact that

$$\langle f, [F] \rangle = \int f(z)\, d\mu(z)$$

where μ is the measure mentioned earlier, and

$$F(w) = \int \frac{1}{w-z}\, d\mu(z).$$

In particular, if $z \in G$ and $F(w) = (z-w)^{-(k+1)}$, then for $f \in H(G)$ we have

$$\left\langle f, \left[\frac{1}{(z-w)^{k+1}}\right]\right\rangle = \frac{1}{2\pi i} \int\limits_{|\zeta-z|=\epsilon} f(\zeta)\frac{d\zeta}{(\zeta-z)^{k+1}}.$$

$$= \frac{f^{(k)}(z)}{k!}.$$

We sketch briefly the proof of this theorem. Clearly, every such function F gives rise to a functional L_F and equivalent functions give rise to the same functional, by the weak Cauchy theorem. And each functional L gives rise to a functional F_L. We must prove that $L_{F_L} = L$ and $F_{L_F} = F$. But

$$L_{F_L}(f) = \frac{1}{2\pi i}\int_\Gamma f(w)F_L(w)\, dw = \frac{1}{2\pi i}\int_\Gamma f(w)\int_K \frac{1}{w-z}\, d\mu(z)\, dw$$

$$= \int_K \frac{1}{2\pi i}\int_\Gamma \frac{f(w)}{w-z}\, dw\, d\mu(z) = \int_K f(z)\, d\mu(z) = L(f)$$

and

$$F_{L_F}(w) = L_F\left(\frac{1}{z-w}\right) = \frac{1}{2\pi i}\int_\Gamma F(z)\frac{1}{z-w}\, dz = F(w).$$

We now prove our theorems in function theory. Let $W = \{w_n\}$ be a sequence of points in G'; the same point may occur more than once, or even infinitely many times. By $R(W)$, we denote the set of rational functions spanned by the functions $1/(w-w_n)$, with the following conventions about multiplicities. If the finite point w_n occurs only finitely many times in the sequence W, then we include only $(w-w_n)^{-1}$, but if it occurs infinitely often, then we include $1/(w-w_n)$, $1/(w-w_n)^2, \ldots$. If $w_n = \infty$ then we include the constant function 1 as well as the function w, and in case ∞ has infinite multiplicity then one includes $1, w, w^2, w^3, \ldots$.

Our version of Runge's theorem on approximation by rational functions with "poles in the holes" seems a little stronger than what can be proved by the classical method of translation of poles.

1.3 Runge's Theorem: If W has at least one limit point in each component of G', then $R(W)$ is dense in $H(G)$.

In particular, if G' is connected (i.e., G is simply connected) and $W = [\infty, \infty, \infty, \ldots]$, this result asserts that every function holomorphic on a simply connected region can be uniformly approximated on compact sets by polynomials. The Cauchy integral theorem follows immediately from this, since the integral of a polynomial around any closed rectifiable curve is 0. Also, for any $G, H(G)$ is separable, since the rational functions with rational coefficients and poles on a suitable countable set form a dense subset of $H(G)$.

Proof: By the Hahn–Banach theorem, it is enough to prove that if L is a continuous linear functional on $H(G)$ such that $L(1/w - w_n) = 0$ for each $w_n \in W$ (recall our convention about multiplicities) then $L = 0$. This reduces to showing that if $F(w_n) = 0$ for each $w_n \in W$, where $[F] \in H_0(G')$ then $F = 0$ since

$$F^{(k)}(w) = k!\left\langle \frac{1}{(z-w)^{k+1}}, [F] \right\rangle \qquad k = 0, 1, 2, \ldots$$

$$F^{(k)}(\infty) = -k!\langle z^{k-1}, [F] \rangle \qquad k = 1, 2, 3, \ldots.$$

But this follows easily, since a function holomorphic on a connected set, that vanishes on a sequence with a limit point in that set, must vanish identically there.

For the purposes of the next theorem, we place on $H_0(G')$ the weak topology as the dual of $H(G)$. The strong topology was studied by Köthe ("Dualität in der Funktionentheorie," *J. Reine u. Angew. Mathe* **191** (1953) pp. 30–49). Our next lemma is equivalent to theorem 13 on p. 39 of that paper, since the weakly convergent sequences and strongly convergent sequences on the dual of a Montel space are the same.

1.4 Lemma: A sequence $\{[F_n]\}$ of elements of $H_0(G')$ is convergent to $[F]$ in the weak topology of $H_0(G')$ as the dual of $H(G)$ if and only if there exist a single open set $A \supseteq G'$, representatives F'_n of $[F_n]$, $n = 1, 2, 3, \ldots$, and a representative F' of $[F]$ such that each F'_n and F' are holomorphic on A and such that F'_n converges uniformly to F' on A.

Proof: We use the uniform boundedness principle in the form that if $\{T_n\}$ is a sequence of continuous linear functionals on the Fréchet space $H(G)$ such that $T(f) = \lim T_n(f)$ exists for each $f \in H(G)$, then $\lim T_n(f) = 0$ as $f \to 0$ uniformly for $n = 1, 2, 3, \ldots$. Now suppose $[F_n]$ converges weakly to $[F]$ in $H_0(G')$, and choose a point $z_0 \in \partial G$. (If $z_0 = \infty$, a slight modification is necessary). Let $T_n(f) = \langle f, F_n \rangle$. Then

$$\frac{F_n^{(k)}(z_0)}{k!} = \frac{1}{2\pi i} \int_{\Gamma} \frac{F(z)}{(z - z_0)^{k+1}} \, dz = T_n\left(\frac{1}{(z - z_0)^{k+1}}\right).$$

By the uniform boundedness principle, we have $|T_n(f)| \leq \sigma \|f\|_K$ for some constant σ and some compact subset K of G, so that

$$\left|\frac{F_n^{(k)}(z_0)}{k!}\right| \leq \sigma\left(\frac{1}{\rho(z_0, K)}\right)^{k+1}$$

where $\rho(z_0, K)$ is the distance from z_0 to K, and hence the power series for F_n around z_0 has a positive radius of convergence that is independent of n. This implies that all the F_n are analytic in some one open set $A_1 \supseteq G'$. An easy extra argument shows that the F_n are uniformly bounded on some slightly smaller open set $A_2 \supseteq G'$, and then a simple argument with normal families shows that the F_n converge uniformly to F on a still smaller open set $A \supseteq G'$.

1.5 Interpolation Theorem: *Let G be an open subset of the complex plane, and let $\{z_n\}$, $n = 1, 2, 3, \ldots$, be a sequence of points of G with no limit point in G. Let $p_1, p_2, p_3, \ldots$, be a sequence of positive integers. Then given any family $\{a_{n,k}\}$ of complex numbers, where $k = 0, 1, \ldots, p_n - 1$ and $n = 1, 2, 3, \ldots$, there is a function f holomorphic on G such that $f^{(k)}(z_n) = a_{n,k}$ for $k = 0, 1, \ldots, p_n - 1$ and $n = 1, 2, 3, \ldots$.*

1.6 Remarks: This result easily implies the Mittag–Leffler theorem. To obtain a meromorphic function with prescribed principal part at a given discrete sequence of points, we need only take $f(z) = g(z)/h(z)$, where g and h interpolate the obvious sequence of values at the required points. For example, near $z = 0$ we want

$$g(z)/h(z) = \frac{a_1}{z} + \frac{a_2}{z^2} + \cdots + \frac{a_l}{z^l} + \varphi(z)$$

where $\varphi(z)$ is analytic. We choose $h(z) = z^l(1 + k(z))$ near $z = 0$, and desire

$$\frac{g(z)}{1 + k(z)} = a_1 z^{l-1} + a_2 z^{l-2} + \cdots + a_l + z^l \varphi(z)$$

Choosing $k(z) = z^l m(z)$, we want only

$$g(z) = a_1 z^{l-1} + a_2 z^{l-2} + \cdots + a_l + z^l \psi(z)$$

near $z = 0$, and this can be done via the interpolation theorem.

1.7 Remark: The Mittag–Leffler theorem easily implies the Weierstrass theorem that there exists a function on G whose zero set is any given discrete set. The idea of the proof is that if $f(z)$ has a zero of multiplicity m at w then $f'(z)/f(z)$ has principal part $m/(z - w)$ at the point w.

1.8 Remark: The interpolation theorem easily implies Helmer's theorem that every finitely generated ideal in $H(G)$ is principal. It is easily reduced to proving that if f and g are in $H(G)$ and have no common zeros, then there are functions α and β in $H(G)$ such that $f\alpha + g\beta = 1$, or $\alpha = (1 - g\beta)/f$. For this to make sense, we need only interpolate the function β properly at the zeros of f.

Proof of the Interpolation Theorem: We denote by S the linear space in $H_0(G')$ spanned by $[1/(z_n - w)^{k+1}]$ for k and n as indicated, and define the linear functional L on S by putting

$$L\left(\left[\left(\frac{1}{z_n - w}\right)^{k+1}\right]\right) = \frac{a_{n,k}}{k!}.$$

We shall prove that S is a closed subspace of $H_0(G')$ and that L is continuous for the topology induced on S by the weak topology on $H_0(G')$. Then by the Hahn–Banach theorem, L may be extended to be continuous on all of $H_0(G')$. Because the dual space of $H_0(G')$ is $H(G)$, by a general theorem in functional analysis, there must exist a function $f \in H(G)$ such that

$$L([F]) = \langle f, [F] \rangle$$

Hence, in view of our earlier remarks, we have

$$\frac{f^{(k)}(z_n)}{k!} = \left\langle f, \left[\left(\frac{1}{z_n - w}\right)^{k+1}\right]\right\rangle = L\left(\left[\frac{1}{z_n - w}\right]^{k+1}\right)$$

$$= \frac{a_{n,k}}{k!}$$

so that f is a suitable interpolating function.

We prove explicitly that L is continuous; the same proof shows also that S is closed. It is enough to show that $L^{-1}(0) = \{[F] \in S: L([F]) = 0\}$ is a closed subspace of $H_0(G')$. But by a corollary of the Banach–Dieudonné theorem (Köthe, "Topologische Lineare Räume," Springer Verlag, Berlin, 1960, p. 275), it is enough to prove that $L^{-1}(0)$ is sequentially closed. To prove this, suppose that $[F_n] \in L^{-1}(0)$, $[F] \in H_0(G')$, and that $\lim [F_n] = [F]$. By our lemma, there is an open set $A \supseteq G'$ and there exist representatives F_n of $[F_n]$ and F of $[F]$ such that

F_n and F are holomorphic on A, with F_n converging uniformly to F on A. Without loss of generality, we suppose that each component of A intersects G'. Hence, F_n must be of the form

$$F_n(w) = \sum_{j=0}^{k_n} \sum_{s=0}^{p_j-1} A_{j,s}^{(n)} \frac{1}{(z_j - w)^{s+1}}$$

for some family $\{A_{j,s}^{(n)}\}$ of complex numbers. However, only a finite number of the z_j are outside A. Hence, only a finite number of the terms $(z_j - w)^{-(s+1)}$ can actually appear in all the F_n together. So we can suppose that $k_n = N$, a constant, for $n = 1, 2, 3, \dots$. Since $F_n \to F$ uniformly on compact subsets of A, we see that

$$A_{s,j} = \lim A_{s,j}^{(n)}$$

exists, and thus

$$F(w) = \sum_{j=0}^{N} \sum_{s=0}^{p_j-1} A_{j,s} \frac{1}{(z_j - w)^{s+1}}.$$

Hence, $[F] \in S$ (proving that S is closed) and

$$0 = \lim_{n \to \infty} L([F_n]) = \lim_{n \to \infty} \sum_{j=0}^{N} \sum_{s=0}^{p_j-1} A_{j,s}^{(n)} a_{j,s}$$

$$= \sum_{j=0}^{N} \sum_{s=0}^{p_j-1} A_{j,s} a_{j,s} = L([F]).$$

Hence, $[F] \in L^{-1}(0)$; $L^{-1}(0)$ is consequently closed. Q.E.D.

Clearly in $H(G)$, every principal ideal is closed.

1.9 Ideal Theorem: *Every closed ideal in $H(G)$ is principal.*

Standard proof: Let I be a closed ideal in $H(G)$ and let $\varphi = $ g.c.d. I, that is, let φ be a holomorphic function that vanishes precisely on the common zeros of all functions in I. Consider $I' = I/\varphi = \{f/\varphi : f \in I\}$. To prove that $I = (\varphi)$ it is enough to prove that $I' = (1)$. Because I is closed, I' is closed. So it is enough to prove that if I is a closed ideal with $z(I) = {}_{\mathrm{def}} \{z : f(z) = 0, \forall z \in I\} = \emptyset$ (counting multiplicities) then $I = H(G)$. Choose a compact set $K \subseteq G$ and then a compact set K', $K \subseteq$ int $K' \subseteq K' \subseteq G$ and then let K'' be the inside of the outside of K' with respect to G, that is, K'' is the union of K' and all the components of the complement of K' that lie in G. For each $z_0 \in K''$, there is an $f_{z_0} \in I$ such that $f_{z_0}(z_0) \neq 0$, and by continuity, $f_{z_0} \neq 0$ throughout a neighborhood of z_0. By compactness, there exist finitely many such functions, say $f_1, f_2, \dots, f_n$ that have no common zeros in K''. By Helmer's theorem, if $f_K = $ g.c.d. $\{f_1, f_2, \dots, f_n\}$, then $(f_1, f_2, \dots, f_n) = (f_K)$. Thus $f_K \in I$, and $f_K(z) \neq 0$ for $z \in K''$. Now $1/f_K \in H(\text{int } K'')$ so by Runge's theorem, for any $\epsilon > 0$ there exists a rational function R_ϵ with poles in comp G (since each component of comp int K'' intersects comp G) such that $\|R_\epsilon - 1/f_K\|_K < \epsilon$ and, hence, $\|R_\epsilon f_K - 1\|_K < \epsilon\|f_K\|_K$. Note that $R_\epsilon \in H(G)$. But $R_\epsilon f_K \in I$ since I is an ideal, and since I is closed, $1 \in I$, Q.E.D.

Functional Analysis Proof: Let I be a closed ideal in $H(G)$. For $f \in H(G)$ denote by $z(f)$ the zero set of f, with multiplicities counted, and by $z(I)$ the intersection of $z(f)$ for those $f \in I$. Given a discrete sequence z of complex numbers, with multiplicity, let $I(z)$ be the ideal consisting of all those $f \in H(G)$ that vanish at least on z. Since, by the Weierstrass theorem, there is a function f whose zero set is precisely z, each ideal $I(z)$ is principal, and to prove our theorem, we need only show that $I = I(z(I))$. We use the Hahn–Banach theorem, and need only prove that if $L \in H(G)'$ and L annihilates I, then L annihilates each $g \in H(G)$ such that $z(g) \supseteq z(I)$. So we let $[\Phi]$ be the element of $H_0(G')$ that corresponds to L. We will show that Φ is a rational function with poles only on $z(I)$ of the correct multiplicities, and the rest follows easily. To prove this assertion about Φ, we shall show that for each $f \in I$, $f\Phi$ has an analytic continuation. Let Γ be the contour in G, mentioned ealier, that winds once around a compact set K outside of which Φ is analytic and let $\Psi = f\Phi$ so that Ψ is analytic in G except possibly on K. Since for each $g \in H(G)$, $fg \in I$, we have

$$0 = L(fg) = \frac{1}{2\pi i} \int_\Gamma g(\zeta) f(\zeta) \Phi(\zeta)\, d\zeta = \frac{1}{2\pi i} \int_\Gamma g(\zeta) \Psi(\zeta)\, d\zeta.$$

From this, we conclude that Ψ has an analytic extension to $H(G)$. For let, for any such cycle Γ, $\tilde{\Psi}_\Gamma(z) = 1/2\pi i \int_\Gamma \Psi(\zeta)/(\zeta - z)\, d\zeta$ for those z that Γ winds around once. By the Cauchy integral theorem, $\tilde{\Psi}_\Gamma$ is independent of Γ, and we shall show that $\tilde{\Psi}_\Gamma(z) = \Psi(z)$ for all suitable z. Given z_0 in $G \setminus K$ let Γ_1 be a cycle that winds once around K and has winding number 0 around z_0, let Γ_2 be a small circle that winds once around z_0 and let $\Gamma = \Gamma_1 + \Gamma_2$. Then

$$\tilde{\Psi}_\Gamma(z_0) = \frac{1}{2\pi i} \int_{\Gamma_1} \frac{\Psi(\zeta)}{\zeta - z_0}\, d\zeta = \frac{1}{2\pi i} \int_{\Gamma_1} \frac{\Psi(\zeta)}{\zeta - z_0}\, d\zeta + \Psi(z_0).$$

Letting

$$A(w) = \frac{1}{2\pi i} \int_{\Gamma_1} \frac{\Psi(\zeta)}{\zeta - w}\, d\zeta$$

for w outside Γ_1, we see that $A(w) = 0$ for $w \in G'$ since $(\zeta - w)^{-1}$ then belongs to $H(G)$, and the same holds for all the derivatives of $A(w)$. By the uniqueness theorem for analytic functions, $A(z_0) = 0$ so that $\tilde{\Psi}_\Gamma(z_0) = \Psi(z_0)$, Q.E.D.

1.10 Derivation pairs: Finally, we consider (following a paper to appear in *Funkcialaj Ekvacioj* **10**: 225–227 (1967).) derivation pairs on $H(G)$. We use the term "derivation pair" to denote a pair of continuous linear functionals L, M on $H(G)$ that satisfy, for all $f, g \in H(G)$ the relation

$$L(fg) = L(f)M(g) + M(f)L(g) \qquad (1.10.1)$$

Each point $\alpha \in G$ gives rise to the obvious derivation pair $L(f) = f'(\alpha)$, $M(f) = f(\alpha)$, but there are derivation pairs essentially different from these, and we find them all here. To begin with, we ignore, throughout, the trivial case $L = 0$.

By the representation of $H_0(G')$ as the dual of $H(G)$, we may write

$$L(f) = \frac{1}{2\pi i} \int_\Gamma f(z)l(z)\, dz$$

$$M(f) = \frac{1}{2\pi i} \int_\Gamma f(z)m(z)\, dz$$

where

$$l(w) = L\left(\frac{1}{w - z}\right) \qquad m(w) = M\left(\frac{1}{w - z}\right)$$

and Γ has winding number 1 around each singular point of l and of m. Using (1.10.1) and the identity

$$\frac{1}{w_1 - z}\frac{1}{w_2 - z} = \frac{1}{w_1 - w_2}\left(\frac{1}{w_2 - z} - \frac{1}{w_1 - z}\right)$$

we get

$$\frac{1}{w_1 - w_2}\{l(w_2) - l(w_1)\} = l(w_1)m(w_2) + l(w_2)m(w_1).$$

On letting w_1 approach $w = w_2$, we get

$$l'(w) = -2l(w)m(w).$$

Applying this above, we get

$$\frac{1}{w_1 - w_2}\{l(w_1) - l(w_2)\} = \frac{l(w_1)}{2}\frac{l'(w_2)}{l(w_2)} + \frac{l(w_2)}{2}\frac{l'(w_1)}{l(w_1)}$$

or

$$l'(w_1)(w_1 - w_2)l(w_2)^2 + l(w_1)^2[l'(w_2)(w_1 - w_2) - 2l(w_2)]$$
$$+ l(w_1)2l(w_2)^2 = 0.$$

Similarly, for any w_3 in G', we have

$$l'(w_1)(w_1 - w_3)l(w_3)^2 + l(w_1)^2[l'(w_3)(w_1 - w_3) - 2l(w_3)]$$
$$+ l(w_1)2l(w_3)^2 = 0.$$

Eliminating $l'(w_1)$ from these equations, and setting $w = w_1$, we see that $l(w)$ has the form

$$l(w) = \frac{1}{Aw^2 + Bw + C}$$

for suitable constants A, B, C that depend on w_2 and w_3. Since $l(\infty) = 0$, we cannot have both $A = 0$ and $B = 0$. Hence, $l(w)$ must have one of the forms

$$l(w) = D\left(\frac{1}{w - \alpha} - \frac{1}{w - \beta}\right)$$

$$l(w) = D\left(\frac{1}{(w - \alpha)^2}\right)$$

$$l(w) = D\left(\frac{1}{w - \alpha}\right)$$

where D is a constant. In each of these cases, we solve for $m(w)$ to get the corresponding representations

$$m(w) = \frac{1}{2}\left(\frac{1}{w - \alpha} + \frac{1}{w - \beta}\right)$$

$$m(w) = \left(\frac{1}{w - \alpha}\right)$$

$$m(w) = \frac{1}{2}\left(\frac{1}{w - \alpha}\right).$$

Therefore we have as the only possible derivation pairs, where E denotes a nonzero constant

$$L(f) = E(f(\alpha) - f(\beta)) \qquad M(f) = \frac{1}{2}(f(\alpha) + f(\beta))$$

$$L(f) = Ef'(\alpha) \qquad M(f) = f(\alpha)$$

$$L(f) = Ef(\alpha) \qquad M(f) = \frac{1}{2}f(\alpha).$$

Since it is easy to check that each of these is indeed a derivation pair, we have completely solved the problem.

The corresponding problem is open when G is an arbitrary Riemann surface, and for functions of several complex variables. Finally, we ask whether the relation (1.10.1) between two linear functionals L and M on $H(G)$ with $L \neq 0$ implies that they must be continuous. This has been answered recently in the affirmative by N. R. Nandakumar in a paper to appear in *Proc. Amer. Math. Soc.*

2. SOME APPROXIMATION THEOREMS

In this section, also unpublished joint work with B. A. Taylor, we prove some approximation theorems that follow by duality from a theorem of C. Rényi on the zeros of the power series coefficients of periodic entire functions. One is a theorem about approximation by polynomials, where the domain of approximation is the integers, and the other is a theorem on completeness of certain sets of entire functions, which will appear shortly in the Journal of the Indian Mathematical Society. We begin with a generalization by P. Erdős and A. Rényi of a theorem stated, but not proved, by Pólya, and then prove the result of C. Rényi, which is certainly interesting in its own right.

2.1 Theorem: [P. Erdős and A. Rényi, "On the number of zeros of successive derivatives of analytic functions," *Acta. Math. Acad. Sci. Hungar*, 7: 125–143 (1956).] *Let N_k be the number of zeros of the k-th derivative $f^{(k)}(z)$ of the entire function $f(z)$ in the closed disc $\{z:|z| \leq 1\}$. Then $\lim\inf\limits_{k \to \infty} N_k/k = 0$.*

Proof: By Jensen's theorem, if $g(z)$ is entire, $g(0) \neq 0$, and $z_1, z_2, z_3, \ldots$ are the zeros of $g(z)$ in the disk $\{z:|z| \leq \rho\}$, then we have

$$\log \frac{\rho^n}{|z_1| |z_2| \cdots |z_n|} = \frac{1}{2\pi} \int_{-\pi}^{\pi} \log \left| \frac{g(\rho e^{i\varphi})}{g(0)} \right| d\varphi.$$

If $N_0(g, r)$ denotes the number of zeros of g in the disk $\{z:|z| \leq r\}$, where $r < \rho$, it follows that

$$N_0(g, r) \log \frac{\rho}{r} \leq \max_{|z|=\rho} \log \left| \frac{g(z)}{g(0)} \right|$$

because

$$\log \frac{\rho^n}{|z_1| |z_2| \cdots |z_n|} = \sum_{j=1}^{n} \log \frac{\rho}{|z_j|} \geq N_0(g, r) \log \frac{\rho}{r}.$$

We require the next result.

2.2 Lemma: If $f(z) = \sum\limits_{k=0}^{\infty} a_k z^k$ is entire, and if for some values of $k \geq 0$, $A \geq 1$ and $B > \rho > 0$ we have

$$|a_{k+j}| < \frac{A|a_k|}{B^j}, \qquad j = 1, 2, 3, \ldots \qquad (2.2.1)$$

then for $|z| = \rho$ we have

$$\left|\frac{f^{(k)}(z)}{f^{(k)}(0)}\right| \leq \frac{A}{(1 - \rho/B)^{k+1}}. \qquad (2.2.2)$$

Proof: From (2.2.1), we have $|a_k| > 0$ and

$$\frac{f^{(k)}(z)}{f^{(k)}(0)} = 1 + \sum_{j=1}^{\infty} \frac{a_{k+j}}{a_k} \frac{(k + 1)(k + 2) \cdots (k + j)}{j!} z^j. \qquad (2.2.3)$$

Now for $|x| < 1$

$$\frac{1}{(1 - x)^{k+1}} = 1 + \sum_{j=1}^{\infty} \frac{(k + 1)(k + 2) \cdots (k + j)}{j!} x^j.$$

Hence,

$$\left|\frac{f^{(k)}(z)}{f^{(k)}(0)} - 1\right| \leq A \left(\frac{1}{(1 - \rho/B)^{k+1}} - 1\right),$$

from which (2.2.2) follows.

Continuing with the proof of Theorem 2.1, we note that since f is entire, we have $|a_n|^{1/n} \to 0$ and so we can, for any $B > 0$ find an infinity of values of k for which

$$|a_{k+j}| \leq \frac{|a_k|}{B^j} \qquad j = 1, 2, 3, \ldots.$$

Indeed, if $\max\limits_{n \geq N} |a_n|^{1/n} = |a_{k_N}|^{1/k_N} < 1/B$, which is true for all large N, then $k = k_N$ works, since

$$|a_{k_N+j}| = (|a_{k_N+j}|^{1/(k_N+j)})^{k_N+j}.$$

From Lemma 2.2 now, with $A = 1$, there are infinitely many values of k with

$$\left|\frac{f^{(k)}(z)}{f^{(k)}(0)}\right| \leq \frac{1}{(1 - \rho/B)^{k+1}} \qquad \text{for } |z| = \rho.$$

We now apply the Jensen theorem with $r = 1$, $\rho > 1$ to get

$$N_k(f(z), 1) \log \rho \leq (k + 1) \log \left(\frac{1}{1 - \rho/B}\right)$$

so that

$$N_k \le \frac{(k+1)}{\log \rho} \log \left(\frac{1}{1 - \rho/B} \right)$$

and thus

$$\liminf \frac{N_k}{k} \le \frac{\log \left(\frac{1}{1 - \rho/B} \right)}{\log \rho}.$$

Holding ρ fixed, say $\rho = 2$, and letting $B \to \infty$, we get the result.

For the rest of our results on gap theorems for entire functions, we refer to C. Rényi, "On a conjecture of Pólya," *Acta. Math. Acad. Sci. Hungar.* **7** (1956): 145–149.

2.3 Lemma: (Pólya, C. Rényi) *Let $f(x) = f^{(0)}(x)$ denote a real function in $C^\infty[a, b]$ and let N_n denote the number of different zeros of $f^{(n)}(x)$ in $[a, b]$. Further, let $Z_a(n)$ and $Z_b(n)$ denote the number of those terms of the sequence $f(a), f'(a), \ldots, f^{(n)}(a)$ and $f(b), f'(b), \ldots, f^{(n)}(b)$, respectively, which are equal to 0, where $n = 0, 1, 2, \ldots$. Then*

$$N_n \ge Z_a(n) + Z_b(n) - n, \qquad n = 0, 1, 2, \ldots \tag{2.3.1}$$

Proof: Our proof is by induction on n. Now (2.3.1) is clear for $n = 0$. Suppose it is true for $n - 1$, that is,

$$N_{n-1} \ge Z_a(n-1) + Z_b(n-1) - (n-1). \tag{2.3.2}$$

Let us put

$$\epsilon_n = \begin{cases} 0 & \text{if } f^{(n)}(a) \ne 0 \text{ and } f^{(n)}(b) \ne 0 \\ 2 & \text{if } f^{(n)}(a) = 0 \text{ and } f^{(n)}(b) = 0 \\ 1 & \text{otherwise.} \end{cases}$$

By Rolle's theorem, $f^{(n)}(z)$ has in (a, b) at least $N_{n-1} - 1$ different zeros. Hence

$$N_n \ge \epsilon_n + N_{n-1} - 1. \tag{2.3.3}$$

But

$$\epsilon_n = Z_a(n) + Z_b(n) - (Z_a(n-1) + Z_b(n-1))$$

and on combining (2.3.2) and (2.3.3) with this fact, we get the desired result.

2.4 Theorem: *Let $f(z)$ be a transcendental entire function. Let $Z_a(n)$ and $Z_b(n)$ denote the number of vanishing coefficients among the first $n + 1$ coefficients of the power series expansion of $f(z)$ around the* *points $z = a$ and $z = b$ $(a \neq b)$, respectively. Then we have*

$$\liminf_{n \to \infty} \frac{Z_a(n) + Z_b(n)}{n} \leq 1.$$

Proof: A simple normalization shows that we may take a and b real, with $a < b$. Also, we may suppose that $f(x)$ is real for real x, for if $f(z) = \sum a_n z^n = \sum (\alpha_n + i\beta_n) z^n$ with α_n, β_n real, and if we let $f^*(z) = \sum \alpha_n z^n$, then

$$Z_a^*(n) \geq Z_a(n) \qquad \text{and} \qquad Z_b^*(n) \geq Z_b(n).$$

Now, from $N_n \geq Z_a(n) + Z_b(n) - n$, we get

$$\liminf \frac{Z_a(n) + Z_b(n)}{n} \leq 1 + \liminf \frac{N_n}{n}.$$

But $\liminf N_n/n = 0$, and our proof is done.

2.5 Corollary: *An entire function cannot have Fabry gaps in its power series expansion around two different points unless it is a polynomial.*

Proof: Otherwise

$$1 \geq \liminf \frac{Z_a(n) + Z_b(n)}{n} \geq \liminf \frac{Z_a(n)}{n}$$

$$+ \liminf \frac{Z_b(n)}{n} = 1 + 1 = 2$$

which is a contradiction.

2.6 Theorem: *Let $f(z)$ be a nonconstant periodic entire function, and let $Z(n)$ denote the number of zero coefficients among the first $n + 1$ coefficients of the power series expansion of f around 0. Then*

$$\liminf_{n \to \infty} \frac{Z(n)}{n} \leq \frac{1}{2},$$

that is, the lower density $D(N)$ of the set N of vanishing coefficients of the power series of a periodic entire function cannot exceed $1/2$.

Remark: The function $\sin z$ shows that the constant $1/2$ cannot be improved.

Proof: Suppose $f(z)$ has period 1. Then

$$\liminf \frac{Z_0(n) + Z_1(n)}{n} \leq 1.$$

But $Z_0(n) = Z_1(n) = Z(n)$ and the result follows.

In the sequel, we shall use the following notation. We let Γ be the collection of all two-sided sequences $b = \{b_k\}$, $k = 0, \pm 1, \pm 2, \ldots$, of complex numbers that grow at most exponentially fast: that is, $|b_k| \leq$ $A \exp(B|k|)$ for some constants A and B. Let Γ_+ be the collection of one-sided sequences $d = \{d_k\}$, $k = 0, 1, 2, \ldots$ of complex numbers that grow at most exponentially fast. If p is any polynomial, then $p|\mathscr{Z}$ (respectively, $p|\mathscr{Z}_+$) belongs to Γ (respectively to Γ_+). We call such elements of Γ and Γ_+ *polynomials*. They are just the restrictions of polynomials to the appropriate sets of integers. Here, $\mathscr{Z}$ is the integers and $\mathscr{Z}_+$ is the nonnegative integers.

By N, we denote any collection of non-negative integers, with 0 adjoined. By P_N (respectively, P_N^+) we denote the collection of all polynomials in Γ (respectively, in Γ_+) with exponents drawn only from N, that is

$$p(x) = \sum a_n x^n; \, a_n = 0 \text{ for } n \notin N.$$

We shall be concerned with finding conditions on N such that P_N be dense in Γ (respectively, that P_N^+ be dense in Γ_+) where Γ and Γ_+ are given suitable natural topologies. Our problem is thus similar to the problem solved by the Szasz–Müntz theorem on polynomial approximation with missing exponents, where the domain of approximation is an interval.

We denote by Γ^* (respectively, Γ_+^*) the collection of all sequences $a = \{a_k\}$, $k = 0, \pm 1, \pm 2, \ldots$ (respectively, $c = \{c_k\}$, $k = 0, 1, 2, \ldots$) such that $\sum |a_k b_k| < \infty$ for each $b \in \Gamma$ (respectively, such that $\sum |c_k d_k| < \infty$ for each $d \in \Gamma_+$). It is easy to verify that $a \in \Gamma^*$ if and only if $|a_k|^{1/k} \to 0$ as $|k| \to \infty$, with a similar statement for Γ_+^*. For each $a \in \Gamma^*$, the mapping $\| \ \|_a : \Gamma \to \mathscr{R}$ of Γ into the real numbers $\mathscr{R}$, given by

$$\|b\|_a = \sum |b_k a_k|$$

is a seminorm on Γ. The collection of all such seminorms, as a varies over Γ^* determines a locally convex topology on Γ, the so-called *normal topology*. A similar procedure gives a corresponding topology on Γ_+, and we shall suppose from now on that Γ and Γ_+ are endowed with these topologies. We stress that throughout this section, we assume that $0 \in N$.

2.7 Theorem: *If N has lower density greater than $1/2$, then P_N is dense in Γ.*

2.8 Theorem: *If N contains a set of even (respectively, odd) integers of positive lower density, then P_N^+ is dense in Γ_+.*

We prove these results by showing that they are equivalent to Theorem 2.6. We require three well-known preliminary results.

2.9 Proposition: (Saks and Zygmund, *Analytic Functions*, p. 361). *There is a linear one-to-one correspondence between* Γ^* *and the collection of all periodic entire functions of period* $2\pi i$, *given as follows. Let* $F(z)$ *be such a function. Then*

$$F(z) = \sum_{-\infty}^{\infty} a_k e^{kz} \tag{2.9.1}$$

where

$$a_k = \frac{1}{2\pi i} \int_0^{2\pi i} F(z) e^{-kz} \, dz. \tag{2.9.2}$$

Furthermore,

$$|a_k|^{1/|k|} \to 0 \text{ as } |k| \to \infty \tag{2.9.3}$$

so that the series in (2.9.1) *converges absolutely and uniformly on every compact set. Conversely, for each sequence* $\{a_k\}$ *such that* (2.9.3) *holds, the expression* (2.9.1) *defines an entire function of period* $2\pi i$.

This result is just the usual result for the Laurent series development of a function that is analytic in an annulus, after an exponential change of variable.

2.10 Proposition: *The pairing of* Γ *and* Γ^* *defined by*

$$\langle b, a \rangle = \sum b_k a_k; \, b \in \Gamma, a \in \Gamma^* \tag{2.10.1}$$

establishes Γ^* *as the topological dual space of* Γ.

2.11 Proposition: *The pairing of* Γ_+ *and* Γ_+^* *defined by*

$$\langle d, c \rangle = \sum d_k c_k; \, d \in \Gamma_+, c \in \Gamma_+^* \tag{2.11.1}$$

establishes Γ_+^* *as the topological dual space of* Γ_+.

Propositions 2.10 and 2.11 are easy to verify directly. Their proofs are given, for example in the book, Köthe, *Lineare Topologische Räume*, p. 424.

We shall now prove that Theorem 2.7 is equivalent to Theorem 2.6. From Proposition 2.9, we know that a periodic entire function of period $2\pi i$ is of the form $F(z) = \sum a_k \exp(kz)$ for some $a = \{a_k\} \in \Gamma^*$. Hence,

$$F^{(n)}(0) = \sum_{k=-\infty}^{\infty} a_k k^n, \qquad n = 0, 1, 2 \ldots,$$

using the convention that $0^0 = 1$. Consequently, Theorem 2.6 is equivalent to the following result.

2.12 Proposition: *Suppose that* N *has lower density greater than*

1/2 and that $0 \in N$. *If* $a = \{a_k\} \in \Gamma^*$ *and if* $\sum_{k=-\infty}^{\infty} a_k k^n = 0$ *for each*

$n \in N$ *then* $a_k = 0$ *for* $k = 0, \pm 1, \pm 2, \dots$.

Now it is easy to see that Theorem 2.7 is also equivalent to Proposition 2.12. For by the Hahn–Banach theorem, P_N is dense in Γ if and only if the only continuous linear functional on Γ that annihilates P_N is the zero functional. By Proposition 2.10 each continuous linear functional L on Γ is of the form $L(b) = L_a(b) = \langle b, a \rangle$ for some $a \in \Gamma^*$. Moreover, the continuous linear functional determined by $a \in \Gamma^*$ annihilates P_N if and only if $\sum a_k k^n = 0$ for each $n \in N$. However, $a \in \Gamma^*$ represents the zero functional if and only if $a_k = 0$ for $k = 0$, $\pm 1, \pm 2, \dots$, and the proof is done.

We shall now show that Theorem 2.8 is equivalent to Theorem 2.6. Our first step is to deduce the following two results from Theorem 2.6.

2.13 Proposition: *If F is an even periodic entire function, and if $F^{(n)}(0) = 0$ for every n in a collection N of even nonnegative integers, with $0 \in N$, such that N has positive lower density, then F must be the null function $F = 0$.*

2.14 Proposition: *If F is an odd periodic entire function, and if $F^{(n)}(0) = 0$ for each n in a collection N of odd nonnegative integers, such that N has positive lower density, then F must be the null function $F = 0$.*

These results follow easily from Theorem 2.6 since the union of the odd (respectively, even) positive integers with a set of even (respectively, odd) positive integers of positive lower density must have lower density exceeding $1/2$. Also, Proposition 2.13 and Proposition 2.14 together imply Theorem 2.6 as we see on writing $F = F_1 + F_2$ where $F_1(z) = \frac{1}{2}(F(z) + F(-z))$ and $F_2(z) = \frac{1}{2}(F(z) - F(-z))$. We now prove that Proposition 2.13 is equivalent to Propositon 2.15 below, and that Proposition 2.14 is equivalent to Proposition 2.16 below.

2.15 Proposition: *If $c = \{c_k\}$ belongs to Γ_+ and if $\sum c_k k^n = 0$ for n in a collection N, with $0 \in N$ of even non-negative integers such that N has positive lower density, then $c_k = 0$ for $k = 0, 1, 2, \dots$.*

2.16 Proposition: *If $c = \{c_k\}$ belongs to Γ_+ and if $\sum c_k k^n = 0$ for n in a collection N of odd positive integers, except that $0 \in N$, such that N has positive lower density, then $c_k = 0$ for $k = 0, 1, 2, \dots$.*

To prove that Propositions 2.13 and 2.15 are equivalent, let $c = \{c_k\}$

belong to Γ_+ and let us define

$$F(z) = \sum_{k=0}^{\infty} c_k(e^{kz} + e^{-kz}).$$

Then F is an even periodic entire function. Furthermore, if n is even and positive, then

$$F^{(n)}(0) = 2 \sum_{k=0}^{\infty} c_k k^n$$

so that the equivalence is clear. A similar proof shows that Propositions 2.14 and 2.16 are equivalent. It remains only to prove that Theorem 2.8 is equivalent to Propositions 2.15 and 2.16 together. This follow from the Hahn–Banach theorem by the same argument as before.

An interesting question is whether the condition on the parity of the elements of N can be dropped from the hypotheses of Theorem 2.8.

We now prove a completeness theorem for entire functions. We designate by δ the differential operator

$$(\delta f)(z) = z \frac{df}{dz}$$

so that $\delta(\sum a_n z^n) = \sum n a_n z^n$. If f is an entire function, we let $\{f_n\}$ be the sequence given by $f_n = \delta^n f$ and defined recursively by

$$f_0 = f, \qquad f_{n+1} = \delta f_n.$$

For example, if $f(z) = \exp z$, then $f_n = P_n f$, where $\{P_n\}$ is a sequence of polynomials—we shall call P_n the nth Stirling polynomial. Thus,

$$P_n(z) = e^{-z}\delta^n(e^z)$$

and we have

$$P_0(z) = 1, P_{n+1}(z) = z[P_n(z) + P'_n(z)]$$

so that

$$P_0(z) = 1, P_1(z) = z, P_2(z) = z^2 + z, P_3(z) = z^3 + 3z^2 + z,$$

$$P_4(z) = z^4 + 6z^3 + 7z^2 + z, P_5(z) = z^5 + 10z^4 + 25z^3 + 15z^2 + z, \text{ etc.}$$

We give here conditions on a set N of non-negative integers that the set $\{f_n : n \in N\}$ be total in the space E of all entire functions, in the topology of uniform convergence on compact sets. In addition, we will give sufficient conditions on N that the set $\{P_n : n \in N\}$ be total in E. We recall that a set of elements of a topological vector space is said to be total if the set of finite linear combinations of its elements is dense in the whole space. Like the preceding results, the present one is a dual theorem to the theorem of Mrs. Rényi on periodic entire functions.

Let E_0 be the space of all entire functions f of exponential type, i.e., functions that satisfy an inequality of the form $|f(z)| \le A \exp(B|z|)$. As is well-known and easy to verify, the normalized power series expansions

$$F(z) = \sum_{k=0}^{\infty} \frac{A_k}{k!} z^k$$

of functions $F \in E_0$ are characterized by the property that $|A_k|^{1/k}$ is bounded. The next result is well known—see for example Köthe's book (p. 424).

2.17 Proposition: *The topological dual space of the space E may be identified with E_0. The pairing defining the duality is given by*

$$\langle f, F \rangle = (F(D)f)(0) \underset{\text{def}}{=} \sum \frac{A_k}{k!} f^{(k)}(0)$$

where $f \in E, F \in E_0, D = d/dz$, and the coefficients A_k are given by the above formula.

We may now prove our completeness results.

2.18 Theorem: *If f is an entire function and if $f^{(n)}(0) \ne 0$ for all $n = 0, 1, 2, \ldots$ and if N is a set of even (respectively, odd) non-negative integers, with $0 \in N$, and having postive lower density, then the set $\{f_n : n \in N\}$ is total in E.*

2.19 Corollary: *Let P_n be the n-th Stirling polynomial. If N satisfies the conditions of the above theorem, then the set $\{P_n : n \in N\}$ is total in E.*

This corollary exhibits explicitly a sequence of polynomials with no more than one of each degree and none of odd degree that is nevertheless total in E.

Proof of Theorem 2.18: By the Hahn–Banach theorem, it is enough to prove that the only continuous linear functional L that annihilates $\{f_n : n \in N\}$ is the zero functional. If L is such an annihilating functional, then there exists a function $F \in E_0$ with the expansion $F(z) = \sum (A_k/k!)z^k$, such that $L(h) = \langle h, F \rangle = \sum (A_k/k!)h^{(k)}(0)$ for each $h \in E$. By hypothesis, $L(f_n) = 0$. We let $c_k = (1/k!) A_k f^{(k)}(0)$. We then have

$$L(f_n) = L(\delta^n f) = \sum_{k=0}^{\infty} c_k k^n,$$

so that $\sum c_k k^n = 0$ whenever $n \in N$. But is is easily seen that $c = \{c_k\}$ is an element of Γ_+^* since $|A_k|^{1/k}$ is bounded and $|f^{(k)}(0)/k!|^{1/k}$ tends to

0 as $k \to \infty$. An appeal to Propositions 2.15 and 2.16 now concludes the proof.

3. SOME UNIQUENESS THEOREMS

We follow here some joint work with B. A. Taylor, entitled *Some uniqueness theorems for analytic functions of one and of several complex variables*, which appeared in *Proc. Cam. Philo. Soc.* **64**: 71–82 (1968). We suppose that f is a function that is analytic on a region G in complex n-space $\mathscr{C}^n$, and that $f^{(m)}(u^{(m)}) = 0$ for each $m = (m_1, m_2, \ldots, m_n)$, $m_i = 0, 1, 2, \ldots$, where

$$f^{(m)}(z) = (D^m f)(z) = \frac{\partial^{m_1 + m_2 + \cdots + m_n} f}{(\partial^{m_1} z_1)(\partial^{m_2} z_2) \cdots (\partial^{m_n} z_n)}(z)$$

Here, $z = (z_1, z_2, \ldots, z_n)$ and $\{w^{(m)}\}$ is a multi-indexed sequence of points in $\mathscr{C}^n$. Under suitable conditions on the distribution of the $\{w^{(m)}\}$, related to the growth of f, it follows that f must vanish identically. This subject has been studied extensively in the case $n = 1$ by many authors, but the study for $n > 1$ has just begun. By a new application of the method of perturbation of a basis in a suitably constructed Banach space, we derive here many of the known results in a unified way, and a number of new results as well. Our method uses a special case of an extension by Arsove of a theorem of Paley and Wiener on the perturbation of a basis in a Banach space. We express the condition of Arsove in the present context, and present some simpler conditions that it follows from. This yields some general theorems on uniqueness which we specialize by means of some calculations with convex functions.

We denote by $\mathscr{C}^n$, $\mathscr{R}^n$, and $\mathscr{Z}^n$ the Cartesian product of n copies of the complex numbers $\mathscr{C}$, the real numbers $\mathscr{R}$, and the integers $\mathscr{Z}$, respectively. By $\mathscr{R}_+^n$ and $\mathscr{Z}_+^n$ we mean the subsets of $\mathscr{R}^n$ and $\mathscr{Z}^n$, respectively, of those points, all of whose coordinates are non-negative. For points, z, w in these spaces, we denote by $z + w$ the usual vector sum of z and w, and by $\langle z, w \rangle$ the scalar product

$$\langle z, w \rangle = z_1 \bar{w}_1 + z_2 \bar{w}_2 + \cdots + z_n \bar{w}_n.$$

We further write

$$\|z\| = \{\langle z, z \rangle\}^{1/2} = \{|z_1|^2 + |z_2|^2 + \cdots + |z_n|^2\}^{1/2}$$

and

$$|z| = (|z_1|, \ldots, |z_n|).$$

For $m = (m_1, m_2, \ldots, m_n) \in \mathscr{L}_+^n$ and $z \in \mathscr{C}^n$, we write

$$z^m = z_1^{m_1} z_2^{m_2} \cdots z_n^{m_n}$$

and

$$m! = m_1! m_2! \cdots m_n!.$$

We place on $\mathscr{L}^n$ the partial ordering $\leq$ where $m \leq m'$ provided that $m_1 \leq m_1', m_2 \leq m_2', \ldots, m_n \leq m_n'$. For later purposes, we also place on $\mathscr{L}_+^n$ the total ordering $\preccurlyeq$ given as follows. On $\mathscr{L}_+ = \mathscr{L}_+^1$ it agrees with the usual ordering. Proceeding by recursion, we suppose that we have ordered $\mathscr{L}_+^{n-1}$ by $\preccurlyeq$. We then define, for $m, m' \in \mathscr{L}_+^n$ that $m \preccurlyeq m'$ if and only if the following holds: first $\sum m_i \leq \sum m_i'$ and then if $\sum m_i = \sum m_i'$ then $\bar{m} \preccurlyeq \bar{m}'$, where $\bar{m}$ and $\bar{m}'$ are the points of $\mathscr{L}_+^{n-1}$ obtained by deleting the first coordinate of m and m', respectively. One should draw a simple diagram to illustrate this ordering at least in the case $n = 2$. We observe that if $m \leq m'$ then $m \preccurlyeq m'$.

We denote by $S_n(\mathscr{C}^k)$ the collection of all "sequences" $a = (a^{(m)})$, indexed by $m \in \mathscr{L}_+^n$ where each $a^{(m)}$ belongs to $\mathscr{C}^k$, say $a^{(m)} = (a_1^{(m)}, \ldots, a_k^{(m)})$. We observe that if

$$f(z) = \sum_{p \geq 0} a^{(p)} z^p$$

then

$$\frac{1}{m!} f^{(m)}(z) = \sum_{p \geq 0} \frac{(m + p)!}{m! p!} a^{(m+p)} z^p.$$

If we are given a sequence $u = (u^{(m)})$ in $S_n(\mathscr{C})$ such that each $u^{(m)}$ is positive, and such that the series $\sum u^{(m)} z^m$ is convergent in a neighborhood of $z = 0$, we denote by U the class of all sequences $a = (a^{(m)})$ with $a \in S_n(\mathscr{C})$, such that $a^{(m)} = O(u^{(m)})$.

Definition: The sequence $w = (w^{(m)}) \in S_n(\mathscr{C}^n)$ is called a sequence of uniqueness for U if the only sequence $b \in U$ such that

$$\sum_{p \geq 0} \frac{(m + p)!}{m! p!} b^{(m+p)} (w^{(m)})^p = 0$$

is the null sequence $b = 0$.

Definition: By $E(U)$, we denote the set of all entire functions f with $f(z) = \sum b^{(p)} z^p$ such that $b = (b^{(p)}) \in U$. The next result is obvious, following our earlier remark.

3.1 Proposition: If $w = (w^{(m)}) \in S_n(\mathscr{C}^n)$ is a sequence of uniqueness for U, and if $f \in E(U)$ and $f^{(m)}(w^{(m)}) = 0$ for each $m \in \mathscr{L}_+^n$, then $f = 0$.

We now turn our attention to finding conditions on w that it be a sequence of uniqueness for U, and we do this by functional analysis. If E is a Banach space with norm $\| \ \|$, we say that a sequence $\{x_1, x_2, x_3, \dots\}$ of elements of E is a basis, provided that each element x of E may be written in a unique way as

$$x = \sum_{j=1}^{\infty} a_j x_j$$

where the a_j are scalars, and the series is convergent in the norm topology of E. If $\{x_1, x_2, x_3, \dots\}$ is a basis for E, we say that the sequence $\{y_1, y_2, y_3, \dots\}$ is *triangular* with respect to $\{x_1, x_2, x_3, \dots\}$ to mean that for each $j = 1, 2, 3, \dots$, the element $y_j - x_j$ is in the closed linear span of

$$\{x_{j+1}, x_{j+2}, x_{j+3}, \dots\}.$$

Now, given a multi-indexed sequence $u = (u^{(m)}) \in S_n(\mathscr{C})$ of positive numbers $u^{(m)}$, we may regard u as that measure on $\mathscr{L}^n_+$ that assigns mass $u^{(m)}$ to the point $m \in \mathscr{L}^n_+$. If $a = a^{(m)} \in L^1(u)$, then

$$\|a\| = \sum |a^{(m)}| u^{(m)}$$

We let $\varepsilon^{(m)}$ denote that element of $L^1(u)$ whose m-th coordinate is 1 and whose other coordinates are 0. We see that $\{\varepsilon^{(m)}\}$ is a basis for $L^1(u)$ if we consider it as indexed by $m \in \mathscr{L}^n_+$ with the total ordering $\preccurlyeq$, which makes $\mathscr{L}^n_+$ order isomorphic to $1, 2, 3, \dots$, the usual order model for the indices of a basis. Since all the series we will encounter are absolutely convergent, the order in which they are summed makes no difference. It is easy to see that $\{\varepsilon^{(m)}\}$ is a basis for $L^1(u)$ with the very special property that

$$\left\| \sum c_m \varepsilon^{(m)} \right\| = \sum |c_m| \, \|\varepsilon^{(m)}\|.$$

For each $w \in \mathscr{C}^n$, we let $\delta^{(m)}(w)$ be that sequence in $S_n(\mathscr{C}^n)$ whose p-th coordinate is 0 unless $p \geq m$, in which case the p-th coordinate is $[p!/(m!(p-m)!)] \, w^{p-m}$. In view of the fact mentioned earlier that $m \leq p$ implies $m \preccurlyeq p$, any sequence $\{\delta^{(m)}(w)\}$ is triangular with respect to the basis $\{\varepsilon^{(m)}\}$.

3.2 Proposition: *Let $U = (u^{(m)}) \in S_n(\mathscr{C})$ be a sequence of positive numbers, and let $w = (w^{(m)}) \in S_n(\mathscr{C}^n)$ be a sequence of elements of C^n such that $\{\delta^{(m)}(w^{(m)})\}$ is a basis for $L^1(u)$. Then w is a sequence of uniqueness for U.*

Proof: We must show that under our hypotheses if $b \in U$ and if

$$\sum_{p\,:0} b^{(m+p)} \frac{(m+p)!}{m!\,p!} (w^{(m)})^p = 0 \qquad (*)$$

for each $m \in \mathscr{L}_+^n$ then $b^{(p)} = 0$ for all $p \in \mathscr{L}_+^n$. We define $\lambda: L^1(u) \to \mathscr{C}$ by $\lambda(a) = \sum b^{(p)} a^{(p)}$ where $a = (a^{(p)}) \in L^1(u)$. Then λ is a continuous linear functional, and the equation (*) above asserts simply that $\lambda(\delta^{(m)}(w^{(m)})) = 0$. Since $\{\delta^{(m)}(w^{(m)})\}$ is a basis for $L^1(u)$, we see that $\lambda(a) = 0$ for each $a \in L^1(u)$. In particular, $b^{(p)} = \lambda(\varepsilon^{(p)}) = 0$, and the proof is done.

We recall that for any sequence $w = (w^{(m)}) \in S_n(\mathscr{C}^n)$, the sequence $\{\delta^{(m)}(w^{(m)})\}$ is triangular with respect to the basis $\{\varepsilon^{(m)}\}$. Our problem is to show that if the $w^{(m)}$ are small, then $\{\delta^{(m)}(w^{(m)})\}$ is a basis, and we do this by applying the next result, due to M. G. Arsove, "Paley–Wiener theorem in linear metric spaces," *Pacific J. Math.* **10**: 365–379 (1960) Theorem 7, p. 374.

Theorem: A. Let $\{x_1, x_2, x_3, \ldots\}$ be a basis of the Banach space E, with the property that for each $x \in E$.

$$\|x\| = \sum_{j=1}^{\infty} |c_j|\, \|x_j\| \text{ if } x = \sum_{j=1}^{\infty} c_j x_j.$$

If $\{y_1, y_2, y_3, \ldots\}$ is triangular with respect to $\{x_1, x_2, x_3, \ldots\}$ and if

$$\limsup_{j \to \infty} \frac{\|y_j - x_j\|}{\|x_j\|} < 1,$$

then $\{y_1, y_2, y_3, \ldots\}$ is also a basis for E.

3.3 Proposition: If

$$\limsup_{\|m\| \to \infty} \sum_{p > 0} \frac{|u_{m+p}|}{|u_m|} \frac{(m+p)!}{m!\,p!} |w^{(m)}|^p < 1$$

then $\{\delta^{(m)}(w^{(m)})\}$ is a basis for $L_1(u)$.

Here, $p > 0$ means, as usual, that $p \geq 0$ but $p \neq 0$.

Proof: We merely apply theorem A, on observing that

$$\frac{\|\delta^{(m)}(w^{(m)}) - \varepsilon^{(m)}\|}{\|\varepsilon^{(m)}\|} = \sum_{p > 0} \frac{|u_{m+p}|}{|u_m|} \frac{(m+p)!}{m!\,p!} |w^{(m)}|^p,$$

further that $\|m\| \to \infty$ is equivalent to $m \in \mathscr{L}_+^n$ tending to infinity for the ordering — and that $\{\delta^{(m)}(w^{(m)})\}$ is triangular with respect to $\{\varepsilon^{(m)}\}$ for the ordering — on $\mathscr{L}_+^n$.

3.4 Proposition: Suppose that u satisfies an inequality

$$\left| \frac{u_{m+p}}{u_m} \right| \leq (R^{(m)})^p \text{ for } m, p \in \mathscr{L}_+^n,$$

where $(R^{(m)}) \in S_n(\mathscr{C}^n)$ and $R^{(m)} = (R_1^{(m)}, \ldots, R_n^{(m)})$ is a point of $\mathscr{R}_+^n$ for each $m \in \mathscr{L}_+^n$. Further, suppose that

$$\limsup_{\|m\| \to \infty} \sum_{i=1}^{n} (m_i + 2)R_i^{(m)}|w_i^{(m)}| < \log 2,$$

where $(w^{(m)})$ is an element of $S_n(\mathscr{C}^n)$, $w^{(m)} = (w_1^{(m)}, \ldots, w_n^{(m)})$. Then $\{\delta^{(m)} (w^{(m)})\}$ is a basis for $L^1(u)$.

Proof: Let

$$I(z) = \prod_{i=1}^{n} \frac{1}{1 - z_i} = \sum_{p \geq 0} z^p$$

where $z = (z_1, \ldots, z_n) \in \mathscr{C}^n$ and $|z_i| < 1, i = 1, 2, \ldots, n$. For $m \in \mathscr{L}_+^n$, we let

$$I_m(z) = \frac{1}{m!}(D^{(m)}I)(z)$$

so that

$$I_m(z) = \prod_{i=1}^{m} \frac{1}{(1 - z_i)^{m_i+1}} = \sum_{p \geq 0} \frac{(m + p)!}{m!\,p!} z^p.$$

For $\|m\|$ sufficiently large, we have $R_i^{(m)}|w_i^{(m)}| < \log 2 < 1$, so that

$$\sum_{p > 0} \left| \frac{u_{m+p}}{u_m} \right| \frac{(m + p)!}{m!\,p!} |w_m|^p \leq \sum_{p > 0} \frac{(m + p)!}{m!\,p!} (R^{(m)})^p |w^{(m)}|^p = I_m(|T^{(m)}|) - 1$$

where

$$T_i^{(m)} = R_i^{(m)}|w_i^{(m)}|, \qquad i = 1, 2, \ldots, n.$$

By the preceding proposition, it is enough to prove that

$$\limsup_{\|m\| \to \infty} \sum_{i=1}^{n} (m_i + 1) \log \frac{1}{1 - T_i^{(m)}} < \log 2.$$

But for each $x > 0$, we have

$$\frac{1}{x} \log \frac{1}{1 - x} = 1 + \frac{1}{2}x + \frac{1}{3}x^2 + \cdots < \frac{1}{1 - x}$$

and for $\|m\|$ large,

$$T_i^{(m)} < \frac{\log 2}{m_i + 2}$$

so that

$$\sum_{i=1}^{n} (m_i + 1) \log \frac{1}{1 - T_i^{(m)}} \leq \sum_{i=1}^{n} (m_i + 1) \frac{T_i^{(m)}}{1 - T_i^{(m)}}$$

$$\leq \sum_{i=1}^{n} \frac{(m_i + 1)T_i^{(m)}}{1 - ((\log 2)/m_i + 2))} = \sum_{i=1}^{n} \frac{(m_i + 1)(m_i + 2)T_i^{(m)}}{m_i + 2 - \log 2}$$

$$< \sum_{i=1}^{n} (m_i + 2)T_i^{(m)},$$

and the result follows.

3.5 Proposition: *Let* $u = (u^{(p)})$ *have the special form*

$$u^{(p)} = \exp\left(-\sum_{i=1}^{n} v_i(p_i)\right), \qquad p = (p_1, \ldots, p_n),$$

where $v_1, \ldots, v_n$ *are convex functions defined for* $0 \leq x < \infty$. *If*

$$\limsup_{\|m\| \to \infty} \sum_{i=1}^{n} (m_i + 2)|w_i^{(m)}| \exp\left(-v_i'(m_i)\right) < \log 2 \qquad (3.5.1)$$

then $\{\delta^{(m)}(w^{(m)})\}$ *is a basis for* $L^1(u)$. *Here,* v_i' *denotes right-hand derivative*

$$v_i'(x) = \lim_{h \to 0+} \frac{v_i(x + h) - v_i(x)}{h}.$$

It is possible that $v_i'(0) = -\infty$, *but otherwise* $v_i'(x) > -\infty$. *In case* $v_i'(0) = -\infty$, $i = (i_1, i_2, \ldots, i_k)$, *then we interpret* (3.5.1) *as saying*

$$\limsup_{\|m\| \to \infty} \sum_{i=1}^{n} (m_i + 2)|w_i^{(m)}| \exp\left(-v_i'(m_i)\right) < \log 2$$

$$(m_i \neq 0, i = i_1, \ldots, i_k). \qquad (3.5.1')$$

Proof: We apply the preceding result, using the estimate

$$\left|\frac{u_{m+p}}{u_m}\right| = \exp\left(-\sum_{i=1}^{n} [v_i(m_i + p_i) - v_i(m_i)]\right) \leq \exp\left(-\sum_{i=1}^{n} p_i v_i^*(m_i)\right),$$

where we take $v_i^*(m_i) = v_i'(m_i)$ unless $m_i = 0$ and $v_i'(0) = -\infty$, in which case we take $v_i^*(m_i) = v_i(1) - v_i(0)$. We choose $R_i^{(m)} = \exp(-v_i^*(m_i))$. Q. E. D.

Now let us apply these results. We let

$$\Phi(r) = \prod_{i=1}^{n} \Phi_i(r_i), \quad r = (r_1, r_2, \ldots, r_n),$$

where the functions $\Phi_i(r_i)$ are continuous nondecreasing functions on $0 \leq r_i < \infty$ such that $\Phi_i(x)/x^k \to \infty$ for each $k > 0$ as $x \to \infty$. If $f: \mathscr{C}^n \to \mathscr{C}$ is an entire function, then we let

$$M(r, f) = \sup_{|z| \leq r} |f(z)| = \sup_{\substack{|z_i| \leq r_i \\ i=1,\ldots,n}} |f(z_1, \ldots, z_n)|.$$

Definition: *We let $E(\Phi)$ be the class of all entire functions f such that there exist constants A and B with $M(r, f) \le A\Phi(Br)$.*

3.1 Theorem: *Given a sequence $(w^{(m)})$, $m \in \mathscr{Z}_+^n$ of elements $w^{(m)}$ of $\mathscr{C}^n$ such that*

$$\lim_{\|m\| \to \infty} \sum_{i=1}^{n} (m_i + 2)|w_i^{(m)}| \exp \{-v_i'(m_i)\} = 0,$$

where v_i is the function defined by

$$v_i(x) = \sup_{-\infty < y < \infty} \{xy - \log \Phi_i(e^y)\},$$

then if $f \in E(\Phi)$ and $f^{(m)}(w^{(m)}) = 0$ for all $m \in \mathscr{Z}_+^n$, it follows that $f = 0$.

Proof: It is easy to verify that each function $v_i(x)$ is convex. We define u as in Proposition 3.5. Let A be a positive number, and define $\bar{u}(A)$ by $\bar{u}(A) = (u^{(p)}(A))$, where

$$u^{(p)}(A) = \inf_{r \in R_+^n} \frac{\Phi(Ar)}{r^p},$$

so that

$$u^{(p)}(A) = u^{(p)}(1) A^{p_1 + \cdots + p_n}.$$

But,

$$u^{(p)}(1) = \exp \left\{-\sum_{i=1}^{p} v_i(p_i)\right\}$$

so that

$$u^{(p)}(A) = \exp \left\{-\sum_{i=1}^{p} v_i(p_i; A)\right\}$$

where

$$v_i(p_i; A) = v_i(p_i) + p_i \log A.$$

Now $E(\bar{u}(A)) \subseteq E(\Phi)$ for all $A > 0$, where $E(\bar{u}(A))$ is defined as in the paragraph preceding Proposition 3.1. But if $f \in E(\Phi)$, then $M(r, f) \le A\Phi(Ar)$ for some $A > 0$, so that then, if $f(z) = \sum a^{(p)} z^p$, then

$$|a^{(p)}| \le \inf \frac{M(r, f)}{r^p} \le Au^{(p)}(A),$$

and thus

$$E(\Phi) = \bigcup_{A > 0} E(\bar{u}(A)).$$

The result will now follow from Proposition 3.5 if we can prove that (3.5.1) holds for each $\bar{u}(A)$. Now (3.5.1) becomes

$$\lim_{\|m\|\to\infty} \sum_{i=1}^{n} (m_i + 2)|w_i^{(m)}|A \exp\{-v_i'(m_i)\}$$

$$= A \limsup_{\|m\|\to\infty} \sum_{i=1}^{n} (m_i + 2)w_i^{(m)} \exp\{-v'(m_i)\} = 0 < \log 2$$

so our proof is complete.

3.6 Proposition (Takenaka): *Let f be an entire function of one complex variable, of growth order at most ρ, type less than τ. If $f^{(m)}(w^{(m)})$ = 0 for $m = 0, 1, 2, \ldots$ where*

$$\limsup_{m\to\infty} m^{1-1/\rho}|w^{(m)}| \leq \frac{\log 2}{\tau\rho^{1/\rho}}$$

then $f = 0$.

Proof: We take

$$v(x) = \frac{x}{\rho} \log \frac{x}{e\sigma\rho},$$

where σ is chosen so that $\sigma < \tau$ and

$$|f(z)| \leq A \exp(\sigma|z|^\rho).$$

Suppose $f(z) = \sum a^{(m)} z^m$. Then

$$|a^{(m)}| \leq A \inf \frac{\exp(\sigma r^\rho)}{r^m}.$$

We choose

$$r = \left(\frac{m}{\rho\sigma}\right)^{1/\rho}$$

to get

$$|a^{(m)}| \leq A \exp\{-v(m)\}.$$

But

$$v'(x) = \frac{1}{\rho}\left(1 + \log \frac{x}{e\sigma\rho}\right)$$

so that

$$\limsup_{m\to\infty} \{(m + 2)|w^{(m)}| \exp(-v'(m))\} = \limsup \{m^{1-1/\rho}|w^{(m)}|(\sigma\rho)^{1/\rho}\}$$

$$\leq (\log 2)(\sigma/\tau)^{1/\rho} \log 2.$$

The result now follows from Propositions 3.5, 3.2, and 3.1.

We omit the proofs of the next two results, which involve only straightforward computations.

3.7 Proposition: *Suppose* $\Phi(r) = r^{\alpha \log r}$. *Then*

$$E(\Phi) = \{f\colon \log M(r, f) \le \alpha(\log r)^2 + O(\log r)\}$$

and the uniqueness condition becomes

$$\lim_{m \to \infty} m|w^{(m)}| \exp\left\{-\frac{m}{2\alpha}\right\} = 0.$$

3.8 Proposition: *Let* $\Phi(r) = \exp r^\rho$. *Then* $E(\Phi)$ *is the collection of entire functions of growth at most order* ρ, *finite type, and the uniqueness condition becomes*

$$\lim_{m \to \infty} m^{1 - 1/\rho}|w^{(m)}| = 0.$$

We also have the next result.

3.9 Proposition: *Let* $\Phi(r) = \exp(\exp r)$. *Then*

$$E(\Phi) = \{f\colon \log \log M(r, f) = O(r)\}$$

and the uniqueness condition becomes

$$\lim_{m \to \infty} \frac{m}{\log m} |w^{(m)}| = 0.$$

Proof: If $\Psi(y) = \log [\Phi(e^y)]$, then a computation shows that $v_i'(x) = y_1$, where $x = \Psi'(y_1) = \exp\{y_1 + e^{y_1}\}$. Thus, $\exp(-v'(x)) = e^{-y_1} \sim [y_1 + e^{y_1}]^{-1} = [\log x]^{-1}$, so that

$$m|w^{(m)}| \exp(-v'(m)) \sim \frac{m}{\log m} |w^{(m)}|.$$

3.10 Proposition: *Let* f *be an entire function of one complex variable such that* $f^{(m)}(w^{(m)}) = 0$ *for* $m = 0, 1, 2, \ldots$, *where* $\lim \sup m|w^{(m)}| < \infty$. *Then* f *is the null function* $f = 0$.

Proof: Choose Φ so that $f \in E(\Phi)$. This is easy to do. Then, if Φ is suitably chosen, we have $\exp\{-v'(m)\} = O(1)$. Hence,

$$\lim m|w^{(m)}| \exp\{-v'(m)\} = 0,$$

and the result follows by Proposition 3.5.

3.11 Proposition (Ålander): *Choose* ρ *with* $0 < \rho < \infty$. *Suppose* f *is of order at most* ρ *and that* $f^{(m)}(w^{(m)}) = 0$ *for a sequence* $w^{(m)}$, $m = 0, 1, 2, \ldots$, *such that*

$$\lim \sup \frac{\log |w^{(m)}|}{\log m} < \frac{1}{\rho} - 1.$$

Then f must be the null function f = 0.

Proof: We choose σ, τ with $\sigma > \tau > \rho$ so that

$$\limsup \frac{\log |w^{(m)}|}{\log m} < \frac{1}{\sigma} - 1 < \frac{1}{\tau} - 1 < \frac{1}{\rho} - 1.$$

Then with the exception of finitely many values of m,

$$m^{1-1/\sigma}|w^{(m)}| < 1$$

and thus

$$\lim m^{1-1/\tau}|w^{(m)}| = 0.$$

But since $\tau > \rho$, it follows that $f \in E(\Phi)$ where $\Phi(r) = \exp(r^\tau)$. Hence, by Proposition 3.8, $f = 0$.

3.12 Proposition: *If f is of finite order and if $f^{(m)}(w^{(m)}) = 0$, where*

$$\limsup_{m \to \infty} \frac{\log |w^{(m)}|}{\log m} \le -1$$

then f is the null function f = 0.

Proof: We just apply the preceding result.

We now turn to entire functions of several complex variables. Proposition 3.13 generalizes Proposition 3.10.

3.13 Proposition: *Suppose $(w^{(m)})$, $m \in \mathscr{Z}_+^n$, is a sequence of points of $\mathscr{C}^n$ such that*

$$\limsup_{||m|| \to \infty} ||w^{(m)}|| \, ||m|| < \infty.$$

If $f: \mathscr{C}^n \to \mathscr{C}$ is an entire function such that $f^{(m)}(w^{(m)}) = 0$ for all m $\in$ $\mathscr{Z}_+^n$, then f must be null function f = 0.

Proof: Choose

$$\Phi(r) = \prod_{i=1}^{n} \Phi_i(r_i), \qquad r = (r_1, r_2, \ldots, r_n)$$

so that f $\in E(\Phi)$. For example, we choose

$$\Phi(r) = \prod_{i=1}^{n} [1 + M(r_i, \ldots, r_i, :f)].$$

We let $v_i(x)$, $i = 1, 2, \ldots, n$ be the functions associated with Φ as in Theorem 3.1. Now $v_i'(x) \to \infty$ as $x \to \infty$. By Theorem 3.1, it is enough to prove that

$$\sum_{i=1}^{n} (m_i + 2)|w_i^{(m)}| \exp(-v_i'(m_i)) \to 0.$$

This is equivalent to proving that if

$$a_m = (m_i + 2)|w_i^{(m)}| \exp(-v_i'(m_i))$$

then $a_m \to 0$ as $m \to \infty$. By hypothesis,

$$|w_i^{(m)}| \le ||w^{(m)}|| = \frac{O(1)}{||m||}$$

so that

$$a_m \le (m_i + 2) \exp(-v_i'(m_i)) \frac{O(1)}{||m||}.$$

If $(m_i + 2) \le ||m||^{1/2}$, then

$$a_m \le \frac{||m||^{1/2} O(1)}{||m||} = o(1),$$

while if $(m_i + 2) \ge ||m||^{1/2}$ then,

$$a_m \le O(1) \exp(-v_i'(||m||^{1/2})) = o(1).$$

Hence, $a_m \to 0$ as $||m|| \to \infty$ and the result is proved. The next result is a generalization of Proposition 3.8.

3.14 Proposition: *Let positive numbers ρ_i, $i = 1, 2, \ldots, n$, be given and let*

$$\Phi(r) = \prod_{i=1}^{n} \exp(r_i^{\rho_i}).$$

If $f \in E(\Phi)$ and if $f^{(m)}(w^{(m)}) = 0$ for all $m \in \mathscr{L}_+^n$, where $w^{(m)}$ is a sequence of elements of $\mathscr{C}^n$ such that

$$\lim_{||m|| \to \infty} \sum_{i=1}^{n} (m_i + 2)^{1 - 1/\rho_i} |w_i^{(m)}| = 0$$

then f must be the null function $f = 0$.

Proposition 3.14 follows directly from the next result, which is a generalization of Proposition 3.6.

3.15 Proposition: *Suppose that $\rho = (\rho_1, \rho_2, \ldots, \rho_n)$ and that $\tau = (\tau_1, \tau_2, \ldots, \tau_n)$, with $\rho_i > 0$, $\tau_i > 0$ for $i = 1, 2, \ldots, n$, and suppose that $f: \mathscr{C}^n \to \mathscr{C}$ is an entire function for which*

$$M(r, f) = O\left(\exp\left(\sum_{i=1}^{n} \sigma_i r_i^{\rho_i}\right)\right)$$

for some $\sigma = (\sigma_1, \sigma_2, \ldots, \sigma_n)$ with $\sigma_i < \tau_i$ for $i = 1, 2, \ldots, n$. If $w^{(m)}$ is a sequence of elements of $\mathscr{C}^n$, $m \in \mathscr{L}_+^n$, such that

$$\limsup_{||m|| \to \infty} \sum_{i=1}^{n} m_i^* |w_i^{(m)}| \le \log 2$$

where

$$m_i^* = (\tau_i \rho_i)^{1/\rho_i}(m_i + 2)^{1-(1/\rho_i)}, i = 1, 2, \ldots, n,$$

and if $f^{(m)}(w^{(m)}) = 0$ *for all* $m \in \mathscr{L}_+^n$, *then* f *must be the null function* $f = 0$.

Proof: We choose positive numbers σ_i, ε_i such that $\sigma_i + \varepsilon_i < \tau_i$ for $i = 1, 2, \ldots, n$ and such that

$$M(r, f) \leq A \exp \left(\sum_{i=1}^{n} \sigma_i r_i^{\rho_i} \right).$$

Then if $f(z) = \sum a_m z^m$, we apply the Cauchy inequality to get

$$|a_m| \leq A \inf_{r \in R_+^n} \frac{\exp(-\sum \sigma_i r_i^{\rho_i})}{r^m}.$$

Hence, setting $r = (r_1, \ldots, r_n)$ where $r_i = (m_i/\sigma_i \rho_i)^{1/\rho_i}$ we get

$$|a_m| \leq A \exp \left(-\sum_{i=1}^{n} \frac{m_i}{\rho_i} \log \frac{m_i}{e\rho_i \sigma_i} \right).$$

We now let, for $i = 1, 2, \ldots, n$,

$$v_i(x) = \frac{x + 2}{\rho_i} \log \frac{x + 2}{e\rho_i(\sigma_i + \varepsilon_i)}.$$

Since $\sigma_i + \varepsilon_i > \sigma_i$, we see for some constant C that

$$\frac{x}{\rho_i} \log \frac{x}{e\rho_i \sigma_i} \geq v(x) + C$$

and thus

$$|a_m| = O \left(\exp \left(-\sum_{i=1}^{n} v_i(m_i) \right) \right).$$

Now

$$v_i'(m_i) = \frac{1}{\rho_i} \log \frac{m_i + 2}{\rho_i(\sigma_i + \varepsilon_i)} + \frac{1}{\rho_i}$$

so that

$$\sum_{i=1}^{n} (m_i + 2)|w_i^{(m)}| \exp(-v_i'(m_i)) = \sum_{i=1}^{n} (m_i + 2)^{1-1/\rho_i}|w_i^{(m)}|[(\sigma_i + \varepsilon_i)\rho_i]^{1/\rho_i}.$$

Since $\sigma_i + \varepsilon_i < \tau_i$, the hypotheses of Proposition 3.5 are satisfied, and on applying that result, we conclude the proof.

An interesting special case of Proposition 3.15 is attained on setting $\rho_i = 1$, and $\tau_i = n^{-1} \log 2$. If $f: \mathscr{C}^n \to \mathscr{C}$ is an entire function such that

$$|f(z_1, z_2, \ldots, z_n)| < A \exp\{\sigma(|z_1| + |z_2| + \cdots + |z_n|)\}$$

then we say that f is of exponential type at most σ.

3.16 Proposition: *If $f: \mathscr{C}^n \to \mathscr{C}$ is an entire function of exponential type less than $n^{-1} \log 2$ and if $f^{(m)}(w^{(m)}) = 0$ for each $m \in \mathscr{L}_+^n$, where $(w^{(m)})$ is a sequence of elements of $\mathscr{C}^n$ such that $|w_i^{(m)}| \leq 1$ for $i = 1, 2, \ldots, n$ for $m \in \mathscr{L}_+^n$, then f must be null function $f = 0$.*

We now study functions supposed analytic only in a polycylinder. In the case $n = 1$, the following result is due to Kakeya and to Takenaka.

3.17 Proposition: *Suppose that the function f is analytic in the polycylinder*

$$\{z: |z_1| < R_1, |z_2| < R_2, \ldots, |z_n| < R_n\}.$$

If $(w^{(m)})$, $m \in \mathscr{L}_+^n$ is a sequence of points in the polycylinder such that

$$\limsup_{||m|| \to \infty} \sum_{i=1}^n (m_i + 2)|w_i^{(m)}| \frac{1}{R_i} < \log 2, \tag{3.17.1}$$

and if $f^{(m)}(w^{(m)}) = 0$ for all $m \in \mathscr{L}_+^n$, then f must be the null function $f = 0$.

Proof: We chose $r = (r_1, r_2, \ldots, r_n)$ such that $r_i < R_i$ for $i = 1, 2, \ldots, n$, and such that

$$\limsup_{||m|| \to \infty} \sum_{i=1}^n (m_i + 2)|w_i^{(m)}| \frac{1}{r_i} < \log 2. \tag{3.17.2}$$

Suppose that $f(z) = \sum a_m z^m$. Then $a_m = O(u_m)$ where $u_m = r^{-m}$. Further,

$$\sum \frac{(m+p)!}{m!p!} a_{m+p}(w^{(m)})^p = \frac{f^{(m)}(w^{(m)})}{m!} = 0.$$

Now

$$\left|\frac{u_{m+p}}{u_m}\right| \leq \frac{1}{r^p}.$$

So by (3.12) and Proposition 3.4, $a_m = 0$ for all $m \in \mathscr{L}_+^n$.

3.18 Corollary: *Suppose that the function f is analytic in the polycylinder*

$$\{z: |z_i| < R; i = 1, 2, \ldots, n\}$$

and suppose that

$$\limsup_{||m||\to\infty} ||m||\, ||w^{(m)}|| < R \log 2.$$

Then if $f^{(m)}(w^{(m)}) = 0$ *for all* $m \in \mathscr{L}_+^n$ *it follows that* f *must be the null function* $f = 0$.

Proof: We will show that (3.17.1) holds. To prove this, notice that we have, by the Schwarz inequality

$$\sum_{i=1}^{n} (m_i + 2)|w_i^{(m)}| \le \left[\sum_{i=1}^{n} (m_i + 2)^2\right]^{1/2} \left[\sum_{i=1}^{n} |w_i^{(m)}|^2\right]^{1/2} = ||m+2||\, ||w^{(m)}||.$$

Since $||m + 2||$ is asymptotic to $||m||$, (3.17.1) follows.

4. VECTOR SPACES OF ENTIRE FUNCTIONS AND DIFFERENTIAL EQUATIONS OF INFINITE ORDER

We present here a brief summary of the main results of the thesis, B. A. Taylor, *Duality and Entire Functions*, University of Illinois, 1965. An expanded version of this thesis appeared in the *Proceedings of the 1966 La Jolla Summer Institute on Entire Functions and Related Parts of Analysis.* Note: Appeared under the title "Some locally convex spaces of entire functions," *Proceedings of Symposia in Pure Mathematics, Vol. II, Entire Functions and Related Parts of Analysis,* Amer. Math. Soc. 1968, pp. 431–467. We consider the differential equation

$$F(D)f = g, \qquad D = \frac{d}{dz} \tag{4.1}$$

where $F(D)$ is a differential operator with constant coefficients, perhaps of infinite order. Here, the functions F, f, and g belong to appropriate spaces of entire functions. For example, if $F(z) = e^z - 1$ then equation 4.1 becomes the difference equation $f(z + 1) - f(z) = g(z)$. We consider two questions here.

1. Is every solution f of the homogeneous equation $F(D)f = 0$ expressible as a linear combination of exponential solutions?

2. Given F and g, does there exist at least one solution f of the equation $F(D)f = g$?

These questions have been studied by numerous mathematicians, for example by L. Ehrenpreis, C. Guichard, B. Malgrange, H. Muggli, J. F. Ritt, H. S. Shapiro, L. Schwartz, P. C. Sikkema. We shall see that the response is affirmative if suitable conditions are satisfied by our spaces of entire functions.

Let C_0 be the space of continuous complex valued functions on the complex plane, that tend to zero at infinity. Let C_0^+ be the set of non-negative functions in C_0, let C_{00} be the set of functions of compact support in C_0, and let C_{00}^+ be the set of non-negative functions in C_{00}. A "set of weights" is a family K of functions k in C_0^+ that satisfy the following conditions:

1. If $k \in K$ and $c > 0$, then $ck \in K$.
   If $k' \in C_0^+, k' \leq k$, then $k' \in K$.
   If $k, k' \in K$, then $\max(k, k') \in K$.
2. $K \supseteq C_{00}^+$.
3. For each $k \in K$, there exists $k' \in K$ such that $k'(z) = k'(|z|)$ and $k'(z) \geq k(z)$.
4. If $k(z) \in K$ and $c \in \mathscr{C}$, then $k(cz) \in K$.
5. If $k \in K$ and $c \in \mathscr{C}$, then $k(z) \exp(cz) \in C_0$.

4.2 Definition: *Let $E(K)$ be the space of all entire functions f such that $kf \in C_0$ for each $k \in K$. We put on $E(K)$ the locally convex topology given by the seminorms.*

$$\|f\|_k = \sup |f(z)k(z)|.$$

4.3 Remark: The condition $kf \in C_0$ for each $k \in K$ is equivalent, for entire functions, to the condition that kf be bounded for each $k \in K$.

Example 1: $K = C_{00}^+$. Then $E(K)$ is the space of all entire fnuctions, under the topology of uniform convergence on compact sets.

Example 2: K is the set of all continuous functions in the plane, whose decrease at ∞ is more rapid than any exponential, that is, $k \in K$ if and only if $k(z) \exp(cz) \in C_0$ for each $c \in \mathscr{C}$. In this case, $E(K) = E_0$, the space of all entire functions of exponential type, in a topology considered by Ehrenpreis and by Malgrange.

We shall see that E and E_0 are dual Montel spaces, and that the topology on E_0 is the Mackey topology with respect to that duality.

4.4 Definition: *Let $M(K)$ be the vector space of all complex Borel measures on $\mathscr{C}$ of the form*

$$d\nu(t) = k(t) \, d\mu(t)$$

for $k \in K$ and μ a bounded complex Borel measure on $\mathscr{C}$.

4.5 Definition: *For $\nu \in M(K)$, let the Laplace transform $\nu^\wedge$ of ν be defined by*

$$\nu^\wedge(z) = \int \exp(zt) \, d\nu(t).$$

It is easy to see that $\nu^\wedge$ is an entire function.

4.6 Definition: *Let $M^\wedge(K)$ be the set of all functions $v^\wedge$ for $v \in$ $M(K)$.*

4.7 Proposition: *$M^\wedge(K)$ is the dual space of $E(K)$, where the duality $\langle E(K), M^\wedge(K)\rangle$ is given by*

$$\langle f, v^\wedge \rangle = \int f\,dv = (v^\wedge(D)f)(0), \quad D = \frac{d}{dz},$$

that is, if $v^\wedge(z) = \sum A_n z^n$, then $\langle f, v^\wedge \rangle = \sum A_n f^{(n)}(0)$, where this series must be absolutely convergent.

For example, if v is a unit point mass at α, then $v^\wedge(z) = e^{\alpha z}$, and $v^\wedge(D)f = \sum (D^n f/n!)\alpha^n = f(z + \alpha)$ by Taylor's formula. Thus, the dual space of $E(K)$ is again a space of entire functions. But it is difficult to work in $M^\wedge(K)$ because the characterization of the functions in $M^\wedge(K)$ is not direct, and because the natural topologies on $M^\wedge(K)$ as the dual of $E(K)$ are not given in a sufficiently concrete form. We shall now resolve these difficulties.

4.8 Proposition: *Let $F(z) = \sum b_n z^n$ be an entire function. Then $F \in M^\wedge(K)$ if and only if there exists a function $k \in K$ such that*

$$n!|b_n| \leq \sup t^n k(t).$$

The "Wiener Tauberian theorem" for $M^\wedge(K)$ may be true.

4.9 Corollary: *If $F \in M^\wedge(K)$ and if F has no zeros, then $(1/F) \in M^\wedge(K)$.*

4.10 Conjecture: *$M^\wedge(K)$ is an algebra.*

It would seem to be difficult to find a direct proof of this assertion. We now identify the equicontinuous subsets of $M^\wedge(K)$. For $k \in K$, let

$$B_k = \{F \in M^\wedge(K) \colon F(z) = \int \exp(zt)k(t)\,d\mu(t); \|\mu\| \leq 1\}$$

4.11 Proposition: *The following assertions are equivalent:*

i) $B \subseteq M^\wedge(K)$ and B is equicontinuous.
ii) There is a $k \in K$ such that $B \subseteq B_k$.
iii) *There exists a function $G(z) \in M^\wedge(K)$ with Taylor series $G(z) = \sum b_n z^n$ such that for each $F \in B$, $F(z) = \sum c_n z^n$ with $|c_n| \leq b_n$.*

4.12 Corollary: *The product E_1, E_2 of two equicontinuous subsets of $M^\wedge(K)$ is again equicontinuous.*

4.13 Proposition: *If a generalized sequence of functions in $M^\wedge(K)$ converges on a dense set in $\mathscr{C}$, and if the functions in that generalized* *sequence form an equicontinuous set, then the generalized sequence converges in the weak topology $w(M^\Lambda(K), E(K))$.*

We now define the dual class of weights K^* to the class K.

4.14 Definition: $K^* = \{k^* \in C_0^+ : k(z)k^*(w) \exp(zw)$ *is bounded for each* $k \in K\}$.

4.15 Proposition: K^* *is a class of weights, $E(K^*)$ is always an algebra, $K \subseteq K^{**}$ and $K^* = K^{***}$.*

4.16 Definition: *K is perfect if $K = K^{**}$.*

4.17 Proposition: *$M^\Lambda(K)$ is a dense subspace of $E(K^*)$ and $M^\Lambda(K^*)$ is a dense subspace of $E(K)$. Further, the Mackey topology $m(M^\Lambda(K), E(K))$ and the strong topology $s(M^\Lambda(K), E(K))$ coincide and are finer than the relative topology $\rho(M^\Lambda(K), E(K^*))$ of $M^\Lambda(K)$ as a subspace of $E(K^*)$.*

We shall find (a) conditions on K such that $M^\Lambda(K) = E(K^*)$; and (b) conditions on K such that $m = \rho$.

4.18 Definition: *K is hypoconvex if, given a function $\lambda(r)$ that is a convex function of $\log r$ on the interval $0 < r < \infty$, such that*

$$k(r) \exp \lambda(r) \in C_0 \qquad \text{for each } k \in K$$

then there exists a function $\lambda_1 \geq \lambda$ such that $\lambda_1(r)$ is a convex function of r and such that $k(r) \exp \lambda_1(r) \in C_0$ for each $k \in K$.

4.19 Theorem: *The following conditions are equivalent.*

i) $\rho = m$

ii) *ρ is an admissible topology with respect to the duality between $M^\Lambda(K)$ and $E(K)$*

iii) $E(K) = M^\Lambda(K^*)$

iv) *K is hypoconvex.*

4.20 Definition: *K has a denumerable base if there exist functions $k_n \in K, n = 1, 2, 3, \ldots,$ such that*

i) $k_1 \leq k_2 \leq \cdots$

ii) *given any $k \in K$, there exists an integer $n > 0$ such that $k(z) = O(k_n(z))$.*

4.21 Definition: *K is sequentially determined if there exist continuous increasing functions $\varphi_n(r)$ for $0 \leq r < \infty$ and $n = 1, 2, 3, \ldots$ such that*

i) $0 \leq \varphi_1(r) \leq \varphi_2(r) \leq \cdots$

ii) $K = \{k \in C_0^+ : k(z) \exp \varphi_n(|z|)$ *is bounded for each $n = 1, 2, 3, \ldots\}$.*

For example, the weights associated with E have a denumerable base, and those associated with E_0 are sequentially determined.

4.22 Theorem: *If K has a denumerable base, then $E(K)$ is metrisable and tonnelé.*

4.23 Theorem: *If K is sequentially determined then $E(K)$ is tonnelé.*

4.24 Theorem: *Suppose K is hypoconvex. Then the following conditions are equivalent.*
i) $E(K)$ *is tonnelé*
ii) $M^{\wedge}(K)$ *is a Montel space for the topology ρ*
iii) $M^{\wedge}(K) = E(K^*)$.

In the terminology of Ehrenpreis, $E(K)$ is analytically uniform if and only if K is hypoconvex and $E(K)$ is tonnelé. We now give a theorem which yields all the classical examples and many others. We consider functions Φ and Ψ, called complementary functions in the sense of Young. That is, they are convex, with $\Phi(0) = \Psi(0) = 0$, and $\Phi(z)$
$$= \int_0^z \varphi(t)\, dt,\ \Psi(z) = \int_0^z \psi(t)\, dt \text{ where } \varphi \text{ and } \psi \text{ are increasing unbound-}$$
ed functions and $\varphi(\psi(t)) = \psi(\varphi(t)) = t$ for all t.

4.25 Theorem: *Let $\varphi_1, \varphi_2, \varphi_3, \ldots$ be convex increasing functions on $[0, \infty[$ with continuous first derivatives, such that $\varphi_n(0) = 0$ and*
a) $\varphi_1 \leq \varphi_2 \leq \varphi_3 \leq \cdots$
b) *for each* $n > 0$, $\varphi_n(2r) = O(\varphi_{n+1}(r))$ *as* $r \to \infty$
c) $\varphi_1(r)/r \to \infty$ *as* $r \to \infty$.
Let K be the set of all continuous non-negative functions $k(z)$ such that $k(z) \exp \varphi_n(|z|)$ is bounded for all $n > 0$. Then K is a set of weights and
i) *K is sequentially determined and hypoconvex.*
ii) *K^* is the set of all continuous positive functions k^* such that there exists an $n > 0$ such that $k^*(z) = O(\exp(-\psi(|z|)))$, where ψ_n is the Young complement of φ_n.*
iii) *K^* has a denumerable base and is hyperconvex (see a later definition).*
iv) *K is perfect.*
Moreover,
1) *$E(K)$ is the set of all entire functions f such that there exists an n such that*
$$M(r, f) = O(\exp(\varphi_n(r))).$$

2) $E(K^*)$ *is the set of all entire functions f such that for each $n > 0$*

$$M(r, f) = O(\exp(\varphi_n(r)))$$

3) $M^\wedge(K^*) = E(K)$ *and* $M^\wedge(K) = E(K^*)$
4) $E(K)$ *and* $E(K^*)$ *are dual Montel spaces.*

To illustrate this result, let $1 < p < \infty$, $1 < p_1 < p_2 < \cdots$, and $p_n \to p$. Let $\varphi_n(r) = r^{p_n}$. Then $E(K)$ is the space of all entire functions of order $< p$, and $E(K^*)$ is the space of all entire functions of order $< q$, where $1/p + 1/q = 1$. The following table exhibits some similar cases.

$E(K)$	$E(K^*)$
f of exponential type	f entire
$\log M(r, f) = o(r \log r)$	$\log \log M(r, f) = O(r)$
$\log M(r, f) = O(r \log r)$	$\log \log M(r, f) = o(r)$
f of order ≤ 1	f of finite order
f of order $< p$, finite type, $p \geq 1$	f of order $\leq q$, zero type $\frac{1}{p} + \frac{1}{q} = 1$
f of order $\leq p$	f of order $< q$

We now consider algebras of entire functions that satisfy some special conditions—the so-called Hadamard algebras.

4.26 Definition: *A topological vector space A of entire functions that is also an algebra under pointwise multiplication $(fg)(z) = f(z)g(z)$, (but it is not supposed that multiplication is continuous) is an Hadamard algebra if*
  i) *A contains all polynomials*
 ii) *if $f \in A$ and $f(a) = 0$ then $(f(z)/(z - a)) \in A$*
iii) *given $g \in A, g \not\equiv 0$, there exists a function $h \in A$ and functions $g_n(z) = P_n(z)e_n(z)$ where P_n is a polynomial and e_n is a unit (i.e., $1/e_n$ is entire) such that $gh/g_n \in A$ and $\lim fgh/g_n = f$ for each $f \in A$.*

For example, if $A = E$, the space of all entire functions, then one may choose $h = 1$ and take for g_n the partial products of the Weierstrass product, while if $A = E_0$, the space of all entire functions of exponential type, then it can be shown that one may choose $h = 1$ and take for g_n the partial products of the Hadamard product.

4.27 Definition: *For an ideal I in A, we define $\mathscr{Z}(I)$ as the set of common zeros, with multiplicities, of the functions in I.*

4.28 Definition: *For a sequence $\mathscr{Z}$ of complex numbers, with multiplicity, we denote by $I(\mathscr{Z})$ the ideal of all the functions $f \in A$ such that $f(z) = 0$ (with multiplicity) for each $z \in \mathscr{Z}$.*

4.29 Theorem: *In an Hadamard algebra, each closed ideal I satisfies*

$$I = I(\mathscr{Z}(I)).$$

For spectral synthesis, it is useful to know if $M^{\wedge}(K)$ in the weak topology $w = w(M^{\wedge}(K), E(K))$ is an Hadamard algebra.

4.30 Theorem: *If $E(K)$ is an algebra it must be an Hadamard algebra. Thus, $E(K^*)$ is always an Hadamard algebra.*

4.31 Theorem: *If K is hypoconvex and if $E(K)$ is tonnelé, then $M^{\wedge}(K), w$ is an Hadamard algebra.*

4.32 Theorem: *If K is perfect, then $M^{\wedge}(K), w$ is an Hadamard algebra.*

4.33 Definition: *K is hyperconvex if, given $k \in K$, there exists a radial function $k' \in K, k' \geq k$, such that $(-\log k'(r))$ is convex and increasing whenever $k'(r) \neq 0$.*

4.34 Theorem: *If K is hyperconvex then $M^{\wedge}(K), w$ is an Hadamard algebra.*

These results, in connection with the earlier results, give elementary conditions that $M^{\wedge}(K), w$ be an Hadamard algebra. The proofs depend on the following theorem which is a generalization to functions of infinite order of the Hadamard factorization theorem. It is proved elsewhere [L. A. Rubel and B. A. Taylor, "A Fourier series method for meromorphic and entire functions," *Bull. Soc. Math. France* **96**: 53–96 (1968)] by using an analysis of the Fourier series for $\log |f(re^{i\theta})|$, without using a Weierstrass or Hadamard product.

4.35 Theorem: (Generalized Hadamard Product): *Suppose that $\lambda(r)$ is continuous, increasing, and positve for $r \geq 0$. Let g be a nonconstant entire functions such that there exist constants A and B such that*

$$\log M(r, g) \leq A\lambda(Br)$$

Then there exist constants C and D and entire functions $h, g_n, (n = 1, 2, 3, \dots)$ and numbers $0 < R_1 < R_2 < \cdots$ such that

i) $g_n(z) = P_n(z)e_n(z)$, where e_n is a zero-free entire function and P_n is a polynomial whose zeros are the zeros of hg in the disc $\{z\colon |z| \leq R_n\}$

ii) $\lim_{n\to\infty} \dfrac{g(z)h(z)}{g_n(z)} = 1$ uniformly on each compact set

iii) $\log M(r, f) \leq C\lambda(Dr)$ if f is any one of the functions $g, h, gh, gh/g_n$, g_n.

We now consider the question of spectral synthesis.

4.36 Definition: *A variety in $E(K)$ is a closed subspace that is invariant under translation.*

4.37 Definition: *We say that spectral synthesis holds for the variety V if V is the variety generated by the exponential monomials $z^j \exp \lambda z$ that belong to V.*

Observe that the set $V = \{f\colon F(D)f = 0\}$ is a variety, and that the question b) posed at the beginning of this chapter is the question of spectral synthesis for this variety V.

4.38 Theorem: *If every closed ideal I in $M^\Lambda(K), w$ satisfies $I = I(\mathscr{Z}(I))$ then spectral synthesis holds for each variety in $E(K)$.*

We have thus found elementary conditions on K, which are verified in all the classical cases, that spectral synthesis holds for each variety in $E(K)$. We have not found a space $E(K)$ in which spectral synthesis does not hold.

We now seek a stronger representation of functions in a variety than as limits of finite linear combinations of exponentials. Incidentally, we say that a function f is mean periodic if the variety it generates is a proper subspace of $E(K)$. It is easy to see, by the Hahn-Banach Theorem, that f is mean periodic if and only if it is a solution of an equation $F(D)f = 0$ for some $F \in M^\Lambda(K)$, $F \neq 0$. We turn now to canonical series for mean periodic entire functions, which is related to problems of interpolation. Let V be a proper variety and let $I = V^\perp = \{F \in M^\Lambda(K)\colon F(D)f = 0$ for all $f \in V\}$. Let $\mathscr{Z} = \mathscr{Z}(I)$, consist of the points $a_0, a_1, a_2, \ldots$ with multiplicities $m_0, m_1, m_2, \ldots$ Then $p(z) = z^j \exp az$ belongs to V if and only if $\exists n$ such that $a = a_n$ and $0 \leq j \leq m_n$. We suppose that $M^\Lambda(K), w$ is an Hadamard algebra. By p, q, we denote exponential monomials.

4.39 Lemma: *Given an exponential monomial $p \in V$, there exists a continuous linear functional $L_p\colon V \to \mathscr{C}$ such that for $q \in V, L_p(q) = \delta_{p,q}$.*

4.40 Definition: *For $f \in V$, the canonical series $\sum_f$ of f is the*
series

$$f \sim \sum_{p \in V} p(z)L_p(f).$$

The canonical series depends only on the function f and not on the
variety V.

4.41 Definition: *By $F(p)$ we denote $\langle F, p \rangle$. Thus, if $p(z) = z^j \exp$
az then $\langle F, p \rangle = F^{(j)}(a)$.*

4.42 Definition: *A V-sequence is a family $c = (c_p), p \in V$ of com-*
plex numbers c_p.

4.43 Definition: *φ is the vector space of all those V-sequences $c =$
(c_p) such that $c_p = 0$ except for a finite number of $p \in V$.*

4.44 Definition: *Let $c = (c_p) \in \varphi$. For $f \in V, f \sim \sum pd_p$, we de-*
fine the c-mean of f by

$$S(f, c) = \sum_{\substack{p \in P \\ p \text{ maximal}}} d_p \sum_{q \leq p} \binom{m(p)}{m(q)} qc_q,$$

*where $m(p)$ is the multiplicity of p, and we write $q \leq p$ if $p(z) = z^j \exp$
az, $q(z) = z^l \exp az$ (same frequency a) and $j \geq l$.*

Thus, $S(f, c)$ is a continuous linear transformation of V into V.
Each $S(f, c)$ is an exponential polynomial, that is, a finite linear combination of exponential monomials.

4.45 Definition: *An effective summability method for V is a se-*
*quence $c(k)$ of elements of φ for $k = 1, 2, 3, \ldots$ such that if $S_k(f) =$
$S[f, c(k)]$ then for each $f \in V, f = \lim S_k(f)$.*

4.46 Theorem: *Suppose $M^\wedge(K), w$ is an Hadamard algebra. If V
is a proper variety, then there exists an effective summability method
for V.*

For the proof, we take a function $G \in V^\perp, G \not\equiv 0$ and then we
take functions, following the generalized Hadamard product, $G_k(z) =$
$G(z)H(z)/P(z)E_k(z)$ and we take

$$c_p(k) = G_k(p).$$

4.47 Definition: *Let S be the set of all V-sequences $b = (b_p)$ such
that there exists a $k \in K$ and an $A > 0$ such that $|b_p| \leq A\|p\|_k$ for
each $p \in V$.*

It is clear that if $F \in M^\wedge(K)$, then the V-sequence $\{F(p)\}$ belongs to S, since

$$F^{(n)}(z) = \int t^n \exp(zt)\, k(t)\, d\mu(t)$$

for some $k \in K$ and for a measure μ with $\|\mu\| \leq 1$.

4.48 Definition: $\mathscr{L}$ *is an interpolation sequence for* $M^\wedge(K)$ *if, given* $b = (b_p) \in S$, *there exists* $F \in M^\wedge(K)$ *such that* $F(p) = b_p$ *for each* $p \in V$. *Here,* $\mathscr{L} = \mathscr{L}(I) = \mathscr{L}(V^\perp)$.

4.49 Theorem: *Suppose that* $M^\wedge(K), w$ *is an Hadamard algebra, and that* K *has a denumerable base. Then the exponential monomials form an absolutely convergent basis for* V *if and only if* $\mathscr{L}$ *is an interpolation sequence for* $M^\wedge(K)$.

4.50 Theorem: *Suppose that* $M^\wedge(K), w$ *is an Hadamard algebra. If the exponential monomials in* V *form an absolutely convergent basis for* V, *then* $\mathscr{L}$ *is an interpolation sequence for* $M^\wedge(K)$. *If* $\mathscr{L}$ *is an interpolation sequence for* $M^\wedge(K)$, *then the canonical series converges absolutely in* $E(K)$ *for each function in* V, *that is, if* $f \in V, f \sim \sum pd_p$, *then* $\sum |d_p| \, \|p\|_k < \infty$ *for each* $k \in K$.

4.51 Theorem: *Suppose* $M^\wedge(K), w$ *is an Hadamard algebra and that* V *is tonnelé. Then* $\mathscr{L}$ *is an interpolation sequence for* $M^\wedge(K)$ *if and only if the canonical series of each function in* V *is absolutely convergent in* $E(K)$.

We consider now the equation $F(D)f = g, g \neq 0$. Recall that the convolution equation $f * \nu = g$ is equivalent to the equation $\nu^\wedge(D)f = g$.

4.52 Proposition: *If* $f \in E(K)$ *and* $\nu \in M(K)$ *then* $(f * \nu)(z)$ *is an entire function.*

4.53 Proposition: *The mapping* $f \to f * \nu, \ \nu \in M(K)$ *is a transformation of* $M^\wedge(K^*)$ *into* $M^\wedge(K^*)$.

4.54 Corollary: *If* K *is hypoconvex then the mapping* $f \to f * \nu, \ \nu \in M(K)$ *is a transformation of* $E(K)$ *into* $E(K)$.

4.55 Theorem: *Let* ν *be a measure in* $M(K), \nu^\wedge \not\equiv 0, Suppose that one of the following hypotheses is satisfied.*
 i) *K is hypoconvex*
 ii) *K is perfect and is sequentially determined*
 iii) *K is hyperconvex and is sequentially determined.*
Then given $g \in E(K)$, *there exists* $f \in E(K)$ *such that* $f * \nu = g$.

4.56 Theorem: *Let v be a measure in $M(K), v^\wedge \not\equiv 0$. Then given $g \in M^\wedge(K^*)$, there exists $f \in M^\wedge(K^*)$ such that $f * v = g$.*

The proofs of these theorems depend on the following "division lemmas." Their proofs use the characteristic function $T(r, f)$ of Nevanlinna and the inequality

$$T(r, f/g) \leq T(r, f) + T(f, g).$$

4.57 Lemma: *Let $\{G_\alpha\}$ be a generalized sequence in $E(K^*)$ such that for $F \in E(K^*)$, $F \not\equiv 0$, we have $FG_\alpha \to 0$ in $E(K^*)$. Then $G_\alpha \to 0$ in $E(K^*)$.*

4.58 Lemma: *Let $E(K)$ be tonnelé and K either hyperconvex or perfect. Let $\{G_n\}$ be a sequence in $M^\wedge(K)$ such that for $F \in M(K)$, $F \not\equiv 0$, we have $FG_n \to 0$ in the topology $w(M^\wedge(K), E(K))$. Then $G_n \to 0$ in this topology.*

5. THE BOUNDED HOLOMORPHIC FUNCTIONS IN THE WEAK-STAR AND STRICT TOPOLOGIES—DEFINITIONS AND PRELIMINARIES

Consider the following result of J. Wolff, "Sur les séries $\sum a_k/(z - \alpha_k)$," *C. R. Acad. Sci. Paris*, **173**: 1057–1058 (1921).

5.1 Theorem: *There exist non-zero constants $A_k, k = 0, 1, 2, \ldots$ such that $\sum |A_k| < \infty$, and complex numbers α_k with $|\alpha_k| < 1$ such that $\sum\limits_{k=0}^{\infty} A_k/(z - \alpha_k) = 0$ for all z with $|z| > 1$.*

Since the proof is short and simple, we include it.

Proof: Let $\{D_k\}$, $k = 1, 2, 3, \ldots$ be a family of disjoint open discs contained in the open unit disc D such that $\sum \lambda(D_k) = \lambda(D)$, where λ is planar area. It is easy to see that such a family of discs exists. Let α_k be the center of D_k and A_k its area. Finally, let

$$f(z) = \sum_{k=1}^{\infty} \frac{A_k}{z - \alpha_k} \text{ for } |z| > 1.$$

From, say, the Cauchy integral formula, we have

$$\frac{A_k}{z - \alpha_k} = \iint\limits_{D_k} \frac{dx\, dy}{z - x - iy}.$$

Thus,

$$f(z) = \sum_{k=1}^{\infty} \iint_{D_k} \frac{dx\, dy}{z - x - iy} = \iint_D \frac{dx\, dy}{z - x - iy} = \frac{\pi}{z}.$$

We define $\alpha_0 = 0$, $A_0 = -\pi$, and the result follows.

Let us put this result in a more general setting. We may rewrite $\sum A_k/(z - \alpha_k) = 0$ as

$$\int \frac{d\mu(w)}{w - z} = 0$$

for all z with $|z| > 1$, where μ is the complex measure consisting of point masses with complex mass A_k at the point α_k for $k = 0, 1, 2, \ldots$. But from this fact about a measure μ, all of whose mass lies in the open unit disc, namely that its "Cauchy transform" vanishes everywhere in the exterior of the unit disc, it follows easily that

$$\int f(w)\, d\mu(w) = 0$$

for each function f that is bounded and analytic in the unit disc since we have

$$\int \frac{d\mu(w)}{w - z} = -\left[\frac{1}{z} \int d\mu(w) + \frac{1}{z^2} \int w\, d\mu(w) + \frac{1}{z^3} \int w^2\, d\mu(w) + \cdots \right]$$

so that

$$\int w^n\, d\mu(w) = 0, \qquad n = 0, 1, 2, \ldots,$$

and, hence,

$$\int P(w)\, d\mu(w) = 0$$

for each polynomial P. But any function f bounded and holomorphic in D can be boundedly approximated there by a sequence of polynomials. That is, there exists a sequence $\{P_n\}$ of polynomials such that $P_n(z)$ is uniformly bounded and $\lim P_n(z) = f(z)$ for each $z \in D$, and consequently the convergence is uniform on compact subsets of D. One way to see this is to take

$$P_n(z) = \frac{S_0(z) + S_1(z) + \cdots + S_n(z)}{n + 1}$$

where $S_j(z)$ is the j-th partial sum of the power series of $f(z)$. Another approach is the following. Suppose we are given a bounded analytic function f and any disc D' whose closure is contained in D. Supposing that $|f(z)| < 1$ for $z \in D$, we show that there exists a polynomial P with $|P(z)| \leq 2$ for $z \in D$ and with $|P(z) - f(z)| < \varepsilon$ for $z \in D'$, where $\varepsilon > 0$ is given. We let D'' be a disc that contains the closure of D', but whose closure is contained in D. Then, letting, for $r < 1$, $f_r(z) = f(rz)$, we see that $|f_r(z) - f(z)| < \varepsilon/2$ for $z \in D''$, and of course, $|f_r(z)| \leq 1$ for $z \in D$, if we choose r sufficiently close to 1. But by Runge's Theorem, $f_r(z)$ can be approximated by polynomials within $\varepsilon/2$ on D'. By the Lebesgue dominated convergence theorem, $\int f d\mu = 0$.

Thus, we are really concerned with measures in the unit disc that annihilate all bounded analytic functions in the unit disc. At this point, there is no reason to restrict our attention to the unit disc, and we shall consider bounded analytic functions on the open set G, and measures, that live in G, that annihilate them.

It is a natural step to pass from here to a detailed study of the duality

$$\langle f, \mu \rangle = \int f d\mu$$

between the bounded analytic functions f on G and the bounded complex measures μ that have all their mass in G. We do this, using a method of balayage, or sweeping, of measures. This is a tool borrowed from potential theory. In summary, we intend to study the weak topology α that the duality induces on the space $B_H(G)$ of all bounded holomorphic functions on G. This is one approach to these chapters on bounded holomorphic functions.

Another approach stems from the notion of bounded convergence of a sequence of bounded holomorphic functions that we mentioned earlier in connection with polynomial approximation. It is natural to study the strongest topology, on the space of bounded holomophic functions, that has these as its convergent sequences. We shall show that this topology coincides with the *strict* topology β in which a net $\{f_\gamma\}$, $\gamma \in \Gamma$, converges to 0 when $\{f_\gamma k\}$ converges uniformly to 0 for each "weight" function k that is continuous on the closure of G and vanishes on the boundary of G. We shall obtain nice results about the topological algebra $\beta(G)$ in many cases, and in particular the case of the disc.

We show that $B_H(G)$ is always the dual space of a certain quotient space of measures, so that α is just the weak-star topology. We prove that the bounded weak-star topology γ always coincides with β.

The two spaces $\alpha(G)$ and $\beta(G)$ are closely related, although different. For example, $\alpha(G)$ and $\beta(G)$ have the same closed subspaces, the same convergent sequences, the same bounded sets, and the same compact sets. In a later section, we will study the problem of polynomial approximation in these topologies. The present sections are largely contained in our paper with A. L. Shields, "The space of bounded analytic functions of a region, "*Ann. Inst. Fourier* (Grenoble) **16**: 235–277 (1966), although parts of it are from other sources.

Definition and Preliminaries

Here, G denotes a connected open set in the complex plane, G^- is its closure, and ∂G is its boundary, both taken with respect to the Riemann sphere. We shall assume that there is a non-constant bounded and analytic function on G, since otherwise the theory of bounded analytic functions would be trivial. We denote by $B_H(G)$ the set (algebra, vector space) of all bounded holomorphic functions on G, without any topology. We write $\|f\|_\infty = \sup\{|f(z)|: z \in G\}$. The bounded analytic functions on G separate the points of G, for if f is a non-constant bounded analytic function and if $z_1, z_2 \in G$, $z_1 \neq z_2$, then if $g(z) = f(z) - f(z_2)$ and $h(z) = (z - z_1)^{-m}g(z)$, where m is the order of the zero of g at z_1, then $h(z_2) = 0$ but $h(z_1) \neq 0$, and $h \in B_H(G)$. Indeed, given distinct points $z_1, z_2, \ldots, z_n$ in G and complex numbers $w_1, w_2, \ldots, w_n$, there is a function $f \in B_H(G)$ such that $f(z_j) = w_j$ for $j = 1, 2, \ldots, n$. To see this, it is enough to produce, for $j = 1, 2, \ldots, n$, a function $f_j \in B_H(G)$ such that $f_j(z_j) = 1$ but $f_j(z_i) = 0$ if $i \neq j$. We construct f_j as $f_j = \varphi_j^1 \varphi_j^2 \cdots \varphi_j^n$, where φ_j^i is constructed as above so that $\varphi_j^i(z_j) = 1$ but $\varphi_j^i(z_i) = 0$ for $i \neq j$, and $\varphi_j^j \equiv 1$.

5.2 Lemma: If G supports a non-constant bounded analytic function, then there is a non-constant bounded analytic function on G that has infinitely many zeros.

Proof: (Wermer) Let f be a non-constant bounded analytic function on G, say $\|f\| = 1$. Let $\{z_n\}$ be a sequence of points in G such that $|f(z_n)| \to 1$, say $f(z_n) \to 1$ without loss of generality, and write $f(z_n) = w_n$. Let $\alpha_n: D \to D$ be the conformal one: one map of the unit disc D onto itself for which $\alpha_n(w_n) = 0$ but such that $\alpha_n(0)$ is "large," where $f(a) = 0$ is a convenient normalization. More precisely, $\alpha_n(0)$ satisfies $|\alpha_1(0)| > 1/2$, $|\alpha_1(0)\alpha_2(0) \cdots \alpha_n(0)| > 1/2$ for $n = 1, 2, 3, \ldots$. We may have to pass to a subsequence of the sequence $\{z_n\}$ to do this, but it is possible by consideration of the hyperbolic distance. Let $g(z) = \lim \alpha_1(f(z))\alpha_2(f(z)) \cdots \alpha_n(f(z))$, where the limit is over a subsequence. The function g has the required properties.

Second Proof: (Rubel and Ryff) Again, we let f be a nonconstant bounded analytic function on G, with $\|f\|_\infty = 1$, say, and we let $\{z_n\}$ be a sequence of points of G such that $|f(z_n)| \to 1$. For a suitable subsequence $\{n_k\}$, the points $w_k = f(z_{n_k})$ form an interpolating sequence for $B_H(D))$ (see K. Hoffman, "Banach Spaces of Analytic Functions" Prentice Hall, New York, 1962) so that there is a bounded analytic function g in the open unit disc D such that $g(w_0) = 1$ but $g(w_n) = 0$ for $n > 1$. The function $h(z) = g(f(z))$ then has infinitely many zeros in G and yet is not identically 0.

We have as yet put no topology on $B_H(G)$. By $H_\infty(G)$ we denote the Banach algebra with $B_H(G)$ as its underlying algebra under the supremum norm $\|f\|_\infty = \sup\{|f(z)|: z \in G\}$. By $M(G)$ we denote the set of all bounded complex-valued Borel measures μ that have all their mass in G. We consider $M(G)$ as a Banach space in the norm $\|\mu\| = |\mu|(G) = \int d|\mu|$. Now $M(G)$ is paired with $B_H(G)$ via the inner product $\langle f, \mu \rangle = \int f d\mu$. There will be measures $\mu \in M(G)$ such that $\langle f, \mu \rangle = 0$ for all $f \in B_H(G)$, and we will discuss them later. If f is continuous in G with $|f| \leq 1$ there, and if $\mu \in M(G)$, then $|\int f d\mu| \leq \int |f| d|\mu| \leq \|\mu\|$, with equality holding in the first inequality if and only if $f d\mu$ is a constant multiple of a positive measure, and with equality holding in the second inequality if and only if $|f| = 1$ a.e. $|\mu|$.

5.3 Proposition: *The space $H_\infty(G)$ is not separable if it contains a non-constant function.*

Proof: As in the proof of the preceding proposition, we produce a sequence of points $\{z_k\}$ in G that is an interpolating sequence for $H_\infty(G)$. Given any sequence ω of zeros and ones, say $\omega = \{\omega_k\}$, let f_ω be a function in $H_\infty(G)$ such that $f_\omega(z_k) = \omega_k, k = 1, 2, 3, \ldots$. Then $\|f_\omega - f_{\omega'}\|_\infty \geq 1$ if $\omega \neq \omega'$. Since there are uncountable many functions f_ω it follows that $H_\infty(G)$ is not separable.

5.4 Proposition: *If $f \in B_H(G)$, then the supremum norm of f is the same as the norm of f when it is regarded as a linear functional on $M(G)$.*

Proof: Let $\|f\|_L = \sup\{|\int f d\mu|: \|\mu\| \leq 1\}$. Then $\|f\|_L \leq \|f\|$ by the above remarks. But let z_0 be a point of G and let ϵ_0 be the unit point mass at z_0. Then $|f(z_0)| = |\int f d\epsilon_0| \leq \|f\|_L$, and so $\|f\|_\infty \leq \|f\|_L$. Q.E.D.

We introduce the equivalence relation $\sim$ into $M(G)$ by defining $\mu \sim v$ to mean that $\int f d\mu = \int f dv$ for all $f \in B_H(G)$. By $N(G)$ we denote the set of those measures μ with $\mu \sim 0$, so that $N(G)$ is a closed linear subspace of $M(G)$. Let $M'(G) = M(G)/N(G)$ be the quotient space, consisting of equivalence clases $[\mu] = \mu + N(G)$ under the equivalence relation $\sim$, with the usual addition and scalar multiplication, under the quotient norm

$$||[\mu]|| = \inf\{||\nu||: \nu \sim \mu\}.$$

5.5 Proposition: min $\{||\nu||: \nu \sim \mu\}$ *exists if and only if there is a measure μ' equivalent to μ so that $a\mu'$ is a positive measure for some constant a.*

Proof: If such a pair a, μ' exists then we may choose a with $|a| = 1$. Now we let f_a be the constant function with the value a. For any $\nu \in N(G)$, we have

$$||[\mu]|| \leq ||\mu'|| = \int f_a d\mu' = \int f_a d(\mu + \nu) \leq ||f_a||_\infty ||\mu + \nu|| = ||\mu + \nu||$$

and hence $||\mu'|| = ||[\mu]||$. For the converse result, we use a later result, that $H_\infty(G)$ is the conjugate Banach space of the Banach space $M'(G)$. Then there exists an $f \in H_\infty(G)$ with $||f||_\infty = 1$ such that $\int f d\mu = ||[\mu]||$. Suppose now that μ' attains the minimum. Then we have

$$||\mu'|| = ||[\mu]|| = \int f d\mu' \leq ||f|| \, ||\mu'|| = ||\mu'||$$

so that by our earlier remark, $f \, d\mu'$ is a positive measure and $|f(z)| = 1$ a.e. $|\mu'|$. In particular, $|f(z)| = 1$ for at least one $z \in G$, which implies that f is constant by the maximum modulus principle for analytic functions. Q.E.D.

Let $K = K(G)$ denote the set of all non-negative continuous functions k defined on G^- such that $k(z) = 0$ for all $z \in \partial G$. It is easy to see that given $\mu_1, \mu_2, \ldots, \mu_n \in M(G)$, there is some $k \in K$ for which

$$\int \frac{1}{|k|} d|\mu_i| < \infty, \qquad i = 1, 2, \ldots, n.$$

6. BOUNDED HOLOMORPHIC FUNCTIONS CONTINUED— THE TOPOLOGIES

6.1 Definition: Let $\alpha(G)$ denote $B_H(G)$ under the weak topology arising from the duality between $B_H(G)$ and $M'(G)$ via the inner product

$$\langle f, [\mu] \rangle = \langle f, \mu \rangle = \int f d\mu.$$

The basic neighborhoods of 0 *thus have the form*

$$N = N(\mu_1, \mu_2, \ldots, \mu_n; \varepsilon) = \{f \in B_H(G): \left| \int f \, d\mu_i \right| < \varepsilon,$$

$$i = 1, 2, \ldots, n\}.$$

This is the weakest topology on $B_H(G)$ in which all the elements of $M'(G)$ are continuous as linear functionals on $B_H(G)$. A net $\{f_\gamma\}$, $\gamma \in \Gamma$ converges to 0 when $\int f_\gamma d\mu$ converges to 0 for each $\mu \in M(G)$. Hence, an α-convergent net is pointwise convergent, as we see if we take μ to be a point mass.

6.2 Definition: $\beta(G)$ *denotes* $B_H(G)$ *under the topology given by the seminorms*

$$\|f\|_k = \sup \{|f(z)k(z)| : z \in G\}$$

where k *ranges over the class* $K(G)$.

The basic neighborhoods of 0 are the sets $\{f: \|f\|_k < \varepsilon\}$, where $k \in K(G)$ and $\varepsilon > 0$. A net $\{f_\gamma\}$, $\gamma \in \Gamma$ converges 0 in this topology if and only if the associated net $\{f_\gamma k\}$, $\gamma \in \Gamma$, converges to 0 uniformly on G for each fixed $k \in K(G)$. In particular, a β-convergent net must converge uniformly on compact subsets of G. We call the β topology also the strict topology. We will see that the β topology coincides with several natural topologies. In particular, it will be shown that the β topology is the strongest topology on $B_H(G)$ that has, as its convergent sequences, precisely those sequences that are boundedly convergent.

6.3 Proposition: *Every* α-*open set is also* β-*open and every* β-*open set is open in* $H_\infty(G)$.

Proof: For the first part, it suffices to show that each basic α-neighborhood of 0 contains a basic β-neighborhood of 0. Let the α-neighborhood be $N(\varepsilon; \mu_1, \mu_2, \ldots, \mu_n)$. Without loss of generality, $\|\mu_i\| \leq 1$ for $i = 1, 2, \ldots, n$. Now we know that there exists a function $k \in K(G)$ for which $\int (1/k)d|\mu_i| < 1$ for $i = 1, 2, \ldots, n$. Let

$$E = E_k = \{f \in B_H(G): \|f\|_k < \varepsilon\}.$$

Then for $f \in E$,

$$\left| \int f \, d\mu_i \right| \leq \int |kf| \frac{1}{k} \, d|\mu_i| < \varepsilon,$$

and hence $E \subseteq N$. For the second part, let $E = E_k$, as above, be a basic β-neighborhood of 0. Let $d = \max \{k(z): z \in G\}$. Since, for $f \in E$, we have $|f(z)| < \varepsilon/k(z)$ for $z \in G$, we see that the H_∞ ball $\{f: \|f\|_\infty < \varepsilon/d\}$ is contained in E.

6.4 Theorem: $\alpha(G)$ and $\beta(G)$ have the same dual space, namely $M'(G)$.

Proof: As noted above, the elements of $M'(G)$ are continuous linear functionals on $\alpha(G)$. Since β is at least as strong as α, they are also continuous on $\beta(G)$. It thus suffices to show that every β-continuous linear functional is given by integration with respect to some measure in $M(G)$. Let $\lambda: \beta(G) \to \mathscr{C}$ be a β-continuous linear functional, so that there is some $k \in K(G)$ such that λ is bounded with respect to the seminorm $\| \ \|_k$. The collection $\{kf\}$, $f \in B_\Pi(G)$ is a vector subspace of the Banach space $C_0(G^-)$ of all continuous functions on G^- that vanish on ∂G, with the supremum norm, and λ is a continuous linear functional on this space. By the Hahn-Banach Theorem, λ has an extension as a bounded linear functional on all of $C_0(G^-)$. From the Riesz representation theorem, it follows that λ can be represented by some measure $\mu \in M(G)$: $\lambda(f) = \int f \, d\mu$ for all $f \in B_\Pi(G)$, and the proof is done.

6.5 Corollary: $\alpha(G)$ and $\beta(G)$ have the same closed linear subspaces and the same closed convex sets.

6.6 Proposition: $\alpha(G)$, $\beta(G)$, and $H_\infty(G)$ have the same bounded sets.

Proof: From Proposition 6.3, it follows that the norm-bounded sets are also β-bounded and α-bounded. In the other direction, it is enough to prove that if S is α-bounded, then S is norm-bounded. Let us regard the elements of S as linear functionals on the Banach space $M(G)$ and apply the uniform boundedness principle. These functionals are point-wise bounded on $M(G)$, since for any measure μ in $M(G)$, we may consider the α-neighborhood U of 0 given by

$$U = \{f \in B_\Pi(G): \left| \int f \, d\mu \right| < 1\}.$$

Since $\mathcal{E}S = \{\mathcal{E}s: s \in S\} \subseteq U$ for some $\mathcal{E} > 0$, we have $|\int f \, d\mu| < 1/\mathcal{E}$ for all $f \in S$. By the uniform boundedness principle, the set S of linear functionals on $M(G)$ is bounded in norm. Now Proposition 5.4 completes the proof.

6.7 Theorem: Let $\{f_\gamma\}$, $\gamma \in \Gamma$ be a uniformly bounded net of functions $f_\gamma \in B_\Pi(G)$. Then following are equivalent.

  i) $\{f_\gamma\}$ is β-convergent to 0
 ii) $\{f_\gamma\}$ is α-convergent to 0
iii) $\{f_\gamma\}$ converges pointwise to 0
 iv) $\{f_\gamma\}$ converges to 0 uniformly on compact subsets of G.

Proof: That i) ⇒ ii) follows from Proposition 6.3. We have remarked already that ii) ⇒ iii). It is a familiar fact from the theory of normal families that iii) ⇒ iv). We prove the result, then, by proving that iv) ⇒ i). We suppose given a function $k \in K(G)$ and a number $\varepsilon > 0$. Without loss of generality, we suppose that $k(z) \leq m$ in G, where $m = \sup\{|f_\gamma(z)| : \gamma \in \Gamma, z \in G\}$. There is a compact set $C \subseteq G$ such that $k(z) < \varepsilon/m$ for $z \in G \setminus C$. Since $\{f_\gamma\}$ converges uniformly on C, there is an index γ_0 such that $\{f_\gamma\} < \varepsilon/m$ on C whenever $\gamma > \gamma_0$. Hence $k(z)|f_\gamma(z)| < \varepsilon$ for all $z \in G$ and $\gamma > \gamma_0$. Q.E.D.

6.8 Corollary: *The α and β topologies agree on bounded sets.*

6.9 Corollary: *The α and β topologies, restricted to any bounded set, are metric.*

Proof: As in the first chapter, the metric is given by

$$\rho(f, g) = \sum_{n=0}^{\infty} 2^{-n} \frac{\|f - g\|_{K_n}}{1 + \|f - g\|_{K_n}}.$$

6.10 Definition: *The γ topology on $B_H(G)$ is the strongest topology that agrees on bounded sets with the topology of uniform convergence on compact subsets of G.*

We shall see later that $B_H(G)$ is always the dual of $M'(G)$, that α is the weak-star topology and that γ is the bounded weak-star topology. We shall also see that the γ topology coincides with the β topology.

6.11 Corollary: *The α and β topologies have the same compact sets, namely the bounded and closed sets.*

Proof: If S is a bounded set in $\alpha(G)$ or $\beta(G)$ then S is a normal family by Theorem 6.6. Hence each sequence of elements of S contains a subsequence that converges uniformly on compact subsets of G. By Theorem 6.7, this implies convergence in $\alpha(G)$ and $\beta(G)$. Since S is metric, this shows that S is compact. For the converse, a compact set in a topological vector space is always closed and bounded.

6.12 Corollary: *The spaces $\alpha(G)$ and $\beta(G)$ have the same convergent sequences, namely the uniformly bounded sequences that converge pointwise to their limit.*

Proof: A bounded and pointwise convergent sequence must be β-convergent by Theorem 6.7, Also, every β-convergent sequence is α-convergent, and every α-convergent sequence is pointwise convergent. It remains to prove that an α-convergent sequence $\{f_n\}$ is bounded, and this follows from the uniform boundedness principle. For if we regard the

f_n as linear functionals on $M(G)$, then they are pointwise bounded since $\lim \int f_n \, d\mu$ exists by hypothesis. They are therefore bounded in the linear functional norm. But we have seen that this norm is equal to the supremum norm. Q.E.D.

6.13 Proposition: *Neither $\alpha(G)$ nor $\beta(G)$ satisfies the first axiom of countability, and consequently neither $\alpha(G)$ nor $\beta(G)$ is a metric space.*

Proof: Let f be a non-constant function in $B_{II}(G)$ with $\|f\|_\infty = 1$, and let

$$S = \{f^n + nf^m : m, n = 1, 2, 3, \dots \}.$$

The sequence f^n converges to 0 in $\alpha(G)$ and $\beta(G)$ by Corollary 6.12. We prove that $0 \in S^-$ but that no sequence of elements of S can converge to 0. Let U be a neighborhood of 0 in either $\alpha(G)$ or $\beta(G)$, respectively and let V be a neighborhood of 0 such that $V + V \subseteq U$. Choose n so large that $f^n \in V$ and then choose m so large that $nf^m \in V$. Then $f^n + nf^m \in U$ so that $0 \in S^-$. On the other hand, suppose that $\{s_k\}$, $k = 1, 2, 3, \dots$, were a sequence of elements of S that were convergent to 0, say

$$s_k = f^{n_k} + n_k f^{m_k}$$

Since the s_k must be uniformly bounded it follows that n_k must be bounded, so that some integer n occurs infinitely often in the sequence $\{n_k\}$. Passing to a subsequence, we have

$$s_k = f^n + nf^{m_k}.$$

If $m_k \to \infty$ then $s_k \to f^n \neq 0$, while if $m_k = m$ for some m and for infinitely many k, then $f^n + nf^m$ is not 0, but is a limit point of $\{s_k\}$. This is a contradiction. Q.E.D.

Despite the previous results, we have the next theorem, for which there are now several proofs.

6.14 Theorem: *The α and β topologies are different.*

Proof: We choose a compact subset C of G that has a non-void interior, and choose a function $k \in K(G)$ such that $k > 0$ on C. Let $E = \{f \in B_{II}(G) : \|f\|_k < 1\}$. The functions in E are uniformly bounded on C. But no α-neighborhood of 0 can have this property. Indeed, given any finite set of measures, there is an $f \in B_{II}(G)$, $f \neq 0$ that is orthogonal to all of them, because we have seen that $B_{II}(G)$ must be an infinite dimensional vector space because the functions in $B_{II}(G)$ separate the points of G. No matter what the number $\varepsilon > 0$, the functions $\{nf\}$, $n = 1, 2, 3, \ldots$, all lie in the α-neighborhood of 0 determined by these measures and this ε. But the functions $\{nf\}$ are not uniformly bounded on C since f cannot vanish on all of C.

6.15 Proposition: *The space $\beta(G)$ is topologically complete while the space $\alpha(G)$ is not.*

Proof: Let $\{f_\gamma\}, \gamma \in \Gamma$ be a Cauchy net in $\beta(G)$ so that if $k \in K(G)$ is given, the net $\{f_\gamma k\}$ is a Cauchy net in the uniform topology and is consequently uniformly convergent. It follows that on each compact subset of G, $\{f_\gamma\}$ is uniformly convergent, so that the limit function f is analytic on G. Also, for each $k \in K(G)$, we have sup $\{|f(z)\,k(z)|:$ $z \in G\} < \infty$, and hence f is bounded, so that $\beta(G)$ is complete.

To prove that $\alpha(G)$ is not complete, we choose a bounded, continuous, and non-analytic function f on G. Let Γ_0 be the family of all finite subsets of $M(G)$, partially ordered by inclusion. Given $\gamma \in \Gamma_0$, say $\gamma = \{\mu_1, \mu_2, \ldots, \mu_n\}$, we choose a function $f_\gamma \in B_{II}(G)$ such that $\int f_\gamma d\mu_i = \int f d\mu_i, i = 1, 2, \ldots, n$. This is possible because $B_{II}(G)$ is finite dimensional. The net $\{f_\gamma\}$ is a Cauchy net in $\alpha(G)$ but does not converge to an element of $\alpha(G)$, so that $\alpha(G)$ is not complete. Q.E.D.

6.16 Proposition: $\beta(G)$ *is a topological algebra. That is, multiplication is jointly continuous in $\beta(G)$.*

Proof: Let $\{f_\gamma\}, \{g_\gamma\}, \gamma \in \Gamma$, be two nets that converge to 0 in $\beta(G)$, and suppose that $k \in K(G)$ is given. Then

$$|f_\gamma g_\gamma k| = |f_\gamma \sqrt{k}\,|\,|g_\gamma \sqrt{k}\,|$$

and since $\sqrt{k} \in K(G)$, it follows that $\{f_\gamma g_\gamma k\}$ converges uniformly to 0 in G, and hence $\{f_\gamma g_\gamma\}$ converges to 0 in $\beta(G)$. Now suppose that $f_\gamma \to f$ and $g_\gamma \to g$ in $\beta(G)$. Then $(f_\gamma - f)(g_\gamma - g) \to 0$ so $(f_\gamma g_\gamma - fg) + f(g - g_\gamma) + g(f - f_\gamma) \to 0$ in $\beta(G)$. But $f(g - g_\gamma) \to 0$ and $g(f - f_\gamma) \to 0$ in $\beta(G)$, as in easy to see, and so $f_\gamma g_\gamma - fg \to 0$. Q.E.D.

6.17 Proposition: (Waelbroeck) $\alpha(G)$ *is not a topological algebra.*

Proof: Suppose multiplication were jointly continuous for the α topology. We claim that for each $\mu \in M(G)$, there exist finitely many $\mu_i, \nu_i \in M(G), i = 1, 2, \ldots, n$, such that

$$\int fg \, d\mu = \sum_{i=1}^{n} \int f d\mu_i \int g \, d\nu_i.$$

For by hypothesis, the mapping $L: \alpha(G) \times \alpha(G) \to \mathscr{C}$ given by $L(f, g)$

$= \int fg \, d\mu$ is a continuous linear functional. Hence there are neighborhoods N_1, N_2 of 0,

$$N_1 = N_1(\mu_1, \mu_2, \ldots, \mu_n; \varepsilon) = \{f \in \alpha(G) \colon \sum \left| \int f \, d\mu_i \right| < \varepsilon\}$$

$$N_2 = N_2(\nu_1, \nu_2, \ldots, \nu_n; \varepsilon) = \{f \in \alpha(G) \colon \sum \left| \int f \, d\nu_i \right| < \varepsilon\}$$

such that $f \in N_1$ and $g \in N_2$ implies that $|\int fg \, d\mu| < 1$. Given $f, g \in \alpha(G)$, we let

$$f' = \frac{\varepsilon f}{\sum |\int f \, d\mu_i|} \qquad \text{and} \qquad g' = \frac{\varepsilon g}{\sum |\int g \, d\nu_i|}.$$

Then $\sum |\int f' \, d\mu_i| = \sum |\int g' \, d\nu_i| = \varepsilon$, so that $|\int f'g' \, d\mu| < 1$ and thus $|\int fg \, d\mu| \leq c(\sum |\int f \, d\mu_i|)(\sum |\int g \, d\nu_i|)$ with $c = 1/\varepsilon$. By a change of notation, we can suppose that $\mu_i = \nu_i$ above, to get

$$\left| \int fg \, d\mu \right| \leq c \left(\sum_{i=1}^{n} \left| \int f \, d\mu_i \right| \right) \left(\sum_{i=1}^{n} \left| \int g \, d\mu_i \right| \right).$$

We now let F be the finite dimensional vector space generated by μ_1, $\mu_2, \ldots, \mu_n$ and let $F^{\perp} = \{f \in \alpha(G) \colon \int f \, d\nu = 0 \text{ if } \nu \in F\}$. Now the bilinear form $\int fg \, d\mu$ vanishes when either f or g lies in $F^{\perp}$. Hence $\int fg \, d\mu$ induces a bilinear form on $(\alpha(G)/F^{\perp}) \times (\alpha(G)/F^{\perp})$. But the dual of $\alpha(G)/F^{\perp}$ is F, since we can ignore the topology because F is finite dimensional. Such a bilinear form can be represented by an element of $F \times F$. In other words, there are elements ρ_i, σ_i of $M(G)$ such that

$$\int fg \, d\mu = \sum_{i=1}^{n} \left(\int f \, d\rho_i \right) \left(\int g \, d\sigma_i \right). \qquad (6.17.1)$$

Now we regard $M(G)$ as an $\alpha(G)$-module with $f \, d\mu$, for $f \in \alpha(G)$ and $\mu \in M(G)$, defined by $\int g(f \, d\mu) = \int fg \, d\mu$. Now we let $\{z_n\}$ be a sequence of points of G such that for each n there exists a function $f_n \in \alpha(G)$ such that $f_n(z_n) \neq 0$ while $f_n(z_m) = 0$ for $m \neq n$. We may take $f_n(z) = f(z)/(z - z_n)^{r_n}$ where f is a function with an infinite set $\{z_n\}$ of zeros, and r_n is the order of the zero of f at z_n. Such a function f exists by Lemma 5.2. From (6.17.1) it follows that each module of the form $\{f \, d\mu \colon f \in B_n(G)\}$ is finite dimensional as a vector space. But let us take

$$\mu = \sum 2^{-n} \varepsilon_{z_n}$$

where ε_z is the unit point mass at z. Then $\{f \, d\mu \colon f \in B_n(G)\}$ would be finite dimensional, which is impossible since the measures $f_n \, d\mu$ are linearly independent for $n = 1, 2, 3, \ldots$. Q.E.D.

7. BOUNDED HOLOMORPHIC FUNCTIONS, CONTINUED—BALAYAGE

Here, we show by balayage, or sweeping, that any measure in $M(G)$ can be replaced by an equivalent measure, as far as its action on $B_H(G)$ goes, that is absolutely continuous. This has important consequences.

7.1 Theorem: *Given a measure μ in $M(G)$, there exists a measure ν in $M(G)$, such that*

$$\nu \sim \mu \qquad and \qquad \|\nu\| \leq \|\mu\|$$

and such that ν is absolutely continuous with respect to Lebesgue measure in the plane.

We denote Lebesgue planar measure by λ, and by $L^1(G)$ the subspace of $M(G)$ consisting of all measures μ that are absolutely continuous with respect to λ, with norm $\|\mu\| = \int d|\mu|$ as before.

Proof: Following a simplifying suggestion by J. L. Doob, we first sweep a point measure and then sweep a general measure, using an integration process. Let ε_w be the unit point mass at the point $w \in G$, let $d = d(w, \partial G)$ denote the distance from w to the boundary of G, let $d' = \min(d, 1)$, and let $a = d'/3$ and $b = 2d'/3$. Then the closed annulus

$$A_w = \{z: a \leq |z - w| \leq b\}$$

is a subset of G, and varies continuously with w in the sense that a and b are continuous functions of w. We define the measure ν_w by

$$\nu_w(E) = \frac{1}{b - a} \int_{t=a}^{t=b} \left(\frac{1}{2\pi i} \int_{|\zeta - w| = t} \chi_E(\zeta) \frac{d\zeta}{\zeta - w} \right) dt$$

for any Borel subset E of G, where χ_E is its characteristic function. Then $f(w) = \int f \, d\nu_w$ for each $f \in B_H(G)$ so that $\nu_w \sim \varepsilon_w$. This is just the Cauchy integral formula averaged over an annulus. A simple estimate also shows that $\|\nu\|_w = 1$. Also, $\nu_w \in L^1(G)$.

Now for any given measure $\mu \in M(G)$, let ν be defined by

$$\nu(E) = \int \nu_w(E) \, d\mu(w) = \int \left(\int \chi_E(z) \, d\nu_w(z) \right) d\mu(w),$$

where E ranges over the Borel subsets of G. For $f \in B_H(G)$,

$$\int f(z) \, d\nu(z) = \int \left(\int f(z) \, d\nu_w(z) \right) d\mu(w) = \int f(w) \, d\mu(w)$$

so that $v \sim \mu$. Also, if φ is any continuous function on G^- that vanishes on ∂G then

$$\left| \int \varphi \, dv \right| \le \left(\int \left(\int |\varphi(z)| d|v_w(z)| \right) d|\mu(w)| \right) \le \|\varphi\|_\infty \|\mu\|$$

where $\|\varphi\|_\infty = \sup \{|\varphi(z)| : z \in G\}$. Hence $\|v\| \le \|\mu\|$.

Finally, we let E be a subset of G such that $\lambda(E) = 0$. Then $v_w(E) = 0$ for each $w \in G$ and, hence, $v(E) = 0$, so that v is absolutely continuous with respect to λ. Q.E.D.

7.2. We recall that $N(G) \subseteq M(G)$ consists of all measures in $M(G)$ orthogonal to $B_H(G)$. Let $N_\lambda(G)$ consists of all measures μ in $L^1(G)$ with $\mu \sim 0$. The inclusion map is a map of $L^1(G)$ into $M(G)$, and it induces a map of the quotient space $L^1(G)/N_\lambda(G)$ into $M'(G)$. By theorem 7.1, this map is an isometry onto.

7.3 Corollary: *The space $L^1(G)/N_\lambda$ is isometrically isomorphic, under the natural correspondence, to $M'(G)$.*

7.4 Corollary: *The space $M'(G)$ is separable.*

This is surprising because $M(G)$ is far from separable.

7.5 Theorem: *The space $H_\infty(G)$ is the conjugate space of the separable Banach space $M'(G)$.*

Proof: It is enough to prove that $H_\infty(G)$ is the conjugate space of $L^1(G)/N_\lambda(G)$. Let $L^\infty(G)$ be the space of equivalence classes, modulo equality almost everywhere with respect to λ, of complex-valued functions on G that are bounded almost everywhere (λ) in the essential supremum norm. Since $B_H(G)$ may be regarded as a linear subspace of $L^\infty(G)$, it is sufficient to show that $B_H(G)$ is weak-star closed in $L^\infty(G)$ as the dual of $L^1(G)$. By the Banach-Dieudonné Theorem, it is enough to prove that $B_H(G)$ is sequentially weak-star closed.

To this end, let $\{f_n\}$ be a sequence of functions in $B_H(G)$, and suppose that $f \in L^\infty(G)$ and that $f_n \to f$ in the weak-star topology of $L^\infty(G)$. Then by definition of that topology, $\int f_m d\mu$ converges to $\int f d\mu$ for each $\mu \in L^1(G)$, so that $\int f_n \, d\mu$ converges to $\int f d\mu$ for each $\mu \in M(G)$, by Theorem 7.1. As we saw in an earlier argument, the f_n must then be uniformly bounded and must converge uniformly on compact subsets of G to a function $f' \in B_H(G)$. Since $\int f' \, d\mu = \int f d\mu$ for each $\mu \in L^1(G)$, it follows that $f' = f$ almost everywhere (λ). Q.E.D.

7.6 Corollary: *The α topology on $B_H(G)$ is precisely the weak-star topology on $H_\infty(G)$ as the dual space of $M'(G)$, and the γ topology is the bounded-weak-star topology.*

A functional analysis proof of most of Theorem 7.5, communicated by L. Waelbroeck, depends on the following result of his ["Duality and the injective tensor product," *Math. Ann.* **163**:122–126(1966)]. It does not quite identify the pre-dual of $H_\infty(G)$ as $M'(G)$.

Theorem A: a) *Let E be a vector space, let X be a convex, balanced, absorbing subset of E, and let $\mathscr{I}$ be a compact topology on X which can be induced by a separated locally convex topology on E. The space E can then be identified with the dual of a Banach space A in such a way that X corresponds to the unit ball of A^*, bicontinuously for $\mathscr{I}$ and the weak-star topology of the unit ball of A^*. If X is metric, then A is separable.*

b) *A is unique up to isometric isomorphism, and isomorphic with the space of those linear forms on E whose restriction to X is continuous for $\mathscr{I}$, with respect to the supremum norm on X.*

c) *Conversely, a Banach space A is isometrically isomorphic with the space of those linear forms on A^* whose restriction to the unit ball of A^* is weak-star continuous, with respect to the supremum norm on the unit ball of A^*.*

For the proof of Theorem 7.5, we choose $E = B_H(G)$, $X = \{f \in B_H(G): \|f\|_\infty \leq 1\}$, and $\mathscr{I}$ as the topology of point-wise convergence. That X is compact follows from the theory of normal families, and we have already seen that X is metric.

7.7 Corollary: *The space $\alpha(G)$ is separable. Indeed, for each subset S of $\alpha(G)$, there is a countable subset $A \subseteq S$ such that each element of S is the limit of a sequence of elements of A. The same assertion holds for $\beta(G)$.*

Proof: This follows for $\alpha(G)$ immediately from Theorem 4 of Chapter 8 of S. Banach, Théorie des Opérations Linéaires. The assertion for $\beta(G)$ follows since $\alpha(G)$ and $\beta(G)$ have the same convergent sequences.

7.8 Corollary: *A linear subspace of $\alpha(G)$ is closed if and only if it is sequentially closed.*

For by the Banach-Dieudonné theorem, the dual of a separable Banach space in the weak-star topology has this property. Since $\alpha(G)$ and $\beta(G)$ have the same closed linear subspaces and the same convergent sequences, this corollary implies that a linear subspace of $\beta(G)$ is closed if and only if it is sequentially closed. We shall soon see a new proof of the striking fact, discovered by Paul Hessler, that *any* subset of $\beta(G)$ is closed if and only if it is sequentially closed. For facts that we shall use about the bounded-weak-star topology, see Dunford and Schwartz, Linear Operators, Vol. 1.

The concluding results of this section are due to Rubel and Ryff, unpublished.

7.9 Theorem: (Rubel and Ryff) *The β topology coincides with the γ topology.*

Proof: We know that γ is at least as strong as β because β agrees with α on bounded sets. We recall that the bounded-weak-star topology is the topology of uniform convergence on norm-null sequences in the pre-dual of the given space, namely in $M'(G)$. Thus, a basic neighborhood of 0 in the γ topology is given by $\{f \in B_H(G): |\int f d\mu_n| < 1, n = 1, 2, 3, \ldots\}$ where $\{[\mu_n]\}$ is any sequence in $M'(G)$ such that $\lim\limits_{n\to\infty} \|[\mu_n]\| = 0$. Our last assertion says that given $k \in K(G)$, there exists a null sequence $\mu_n \in M'(G)$ such that $|\int f d\mu_n| < 1$ for $n = 1, 2, 3, \ldots$ implies $\sup |f(z)k(z)| < 1$. It would be good to find a direct proof of this. The converse, that the β topology is at least as strong as the γ topology, reduces to proving that, given a norm-null sequence $\{\mu_n\}$ in $M'(G)$, there exists a $k \in K(G)$ such that $\sup |fk| < 1$ implies $|\int f d\mu_n| < 1$, for $n = 1, 2, 3, \ldots$. E. Speer has simplified our original proof of this fact. We construct a sequence $C_m \subseteq G$ of compact sets $C_m, m = 1, 2, 3, \ldots$, with $C_m \subseteq C_{m+1}$ and $\cup C_m = G$, such that

$$|\mu_n|(G \setminus C_m) \leq 2^{-m} \qquad \text{for } n \leq m.$$

We now choose $k \in K$ such that

$$k(z) \geq 4 \max \left(\|\mu_m\|, \frac{1}{m} \right)$$

for $z \in C_m$. Then if $|fk| < 1$ on G, we have

$$\left| \int_G f d\mu_n \right| \leq \int \frac{d|\mu_n|}{k} = \left\{ \int_{C_n} + \int_{C_{n+1}\setminus C_n} + \cdots \right\} \frac{d|\mu_n|}{k}$$

$$\leq \int_{C_n} \frac{d|\mu_n|}{4\|\mu_n\|} + \sum_{m=n}^{\infty} \int_{C_{m+1}\setminus C_m} \frac{m+1}{4} d|\mu_n|$$

$$\leq \frac{1}{4} \left\{ 1 + \sum_{m=n}^{\infty} \frac{m+1}{2^m} \right\} \leq 1.$$

7.10 Corollary: *A subset of $\beta(G)$ is closed if and only if it is sequentially closed.*

Proof: The proof depends on the Banach-Dieudonné Theorem

applied to the bounded-weak-star topology γ on the dual of a separable Banach spach E, and seems to be generally known. It is a consequence of that theorem (see Köthe, "Topologische Lineare Räume," p. 275, (4)) that a subset $M \subseteq E'$ is γ-closed if and only if, for every weakly closed and bounded subset $B \subseteq E'$, it follows that $M \cap B$ is weakly closed. Now if E is separable then (again Köthe, p. 261, (4)) every such set $M \cap B$ is metrisable in the weak topology. Now if M is sequentially closed in the γ topology then since B is weakly closed, it is γ-closed since the γ topology is stronger than the weak-star topology. Hence B is γ-sequentially closed, so $M \cap B$ is γ-sequentially closed and hence weakly sequentially closed, and hence by metrisability, $M \cap B$ is weakly closed, and consequently M is γ-closed.

In the next theorem, the case $G = D$ was first proved by Conway.

7.10 Theorem: *If $G = D$, or more generally, if there is a boundary continuum that is isolated from all the other components of the boundary of G, then the β-topology is different from the Mackey topology.*

Proof: We first require the following lemma on Banach spaces, whose proof we omit.

7.11 Lemma: *For a Banach space B, the bounded weak-star topology on B^* coincides with the Mackey topology if and only if every weakly convergent sequence in B is norm convergent.*

Returning to the proof of the theorem we exhibit a sequence $\{[\mu_n]\}$ in $M'(D)$ that is weakly convergent to 0 but such that $\|[\mu_n]\| \geq \varepsilon > 0$. We let

$$\int f \, d\mu_n = a_n(f) = \frac{f^{(n)}(0)}{n!}.$$

By the Cauchy Integral Theorem, it is easy to realize μ_n as an element of $M'(D)$. Now $\|[\mu_n]\| \geq 1$ since $a_n(z^n) = 1$. But also, $\int f \, d\mu_n \to 0$ for each $f \in B_H(D)$. One way to see this is that $H_\infty \subseteq H_2$ so that for $f \in H_\infty$, $\sum |a_n(f)|^2 < \infty$ so that consequently $a_n \to 0$.

For the more general case we may, by conformal mapping, suppose the distinguished boundary component to be the unit circle, and the other boundary components to lie in the disc $\{|z| < \rho\}$. We now choose σ with $\rho < \sigma < 1$ and let

$$\int f \, d\mu_n = \frac{1}{2\pi i} \int_{|z|=\sigma} \frac{f(z) \, dz}{z^n \, z}.$$

Then $\mu_n(z^n) = 1$ but $\mu_n(f) \to 0$ as $n \to \infty$ by the Riemann-Lebesgue Lemma. The most general case seems not to be easy.

7.12 Proposition: *There is a sequentially closed subset of $\alpha(G)$ that is not closed.*

The proof depends directly on the following theorem in functional analysis.

7.13 Proposition: *Let B be a Banach space such that in the weak-star topology every sequentially closed subset of B^* is closed. Then B is finite dimensional.*

Proof: Suppose B is infinite dimensional. We shall then show that there exists a subset $X \subseteq B^*$ that is closed in the bounded-weak-star topology but that is not weak-star closed. Then X is sequentially closed in the weak-star topology, for by the uniform boundedness principle, X must be norm bounded. But convergence in the weak-star and bounded-weak-star topologies is the same in bounded subsets, by definition. To construct such a set X, we recall that the bounded-weak-star topology is given by uniform convergence on null sequences in B. Given $\{x_n\}$ in B, $x_n \to 0$, the set $\{x' \in B^* : |\langle x', x_n \rangle| < 1\}_{n=1}^{\infty}$ defines a neighborhood of the origin in the bounded-weak-star topology. This neighborhood contains no weak-star neighborhood of the origin if the x_n are suitably chosen. Otherwise, we could choose $y_1, \ldots, y_n \in B$ such that

$$\{x' \in B^* : |\langle x', y_j \rangle| < 1, j = 1, 2, \ldots, n\}$$
$$\subseteq \{x' \in B^* : |\langle x', x_m \rangle| < 1, m = 1, 2, \ldots\}.$$

If B is infinite dimensional, we may assume that the sequence $\{x_n\}$ consists of linearly independent vectors. By the Hahn-Banach theorem, we may choose $x' \in B^*$ so that $|\langle x', y_j \rangle| = 0$, $j = 1, 2, \ldots, n$ but so that $\langle x', x_m \rangle \neq 0$ for at least one m. Then $\lambda x'$ lies in the weak-star neighborhood for all values of λ, but obviously cannot be an element of the bounded-weak-star neighborhood, if we choose λ large. Hence one may take for X the complement of the bounded-weak-star neighborhood.

7.14 Theorem: *The β topology on $B_{\Pi}(G)$ is the strongest topology in which a sequence converges to its limit if and only if it is uniformly bounded and pointwise convergent to that limit.*

This clearly follows from the next result.

7.15 Theorem: *Let B be a separable Banach space and B^* its dual. Denote by α the weak-star topology on B^* and by γ the bounded-weak-star topology on B^*. Then γ is the strongest topology on B^* that has the same convergent sequences (with the same limits, of course) as α.*

Proof: The proof depends on some results of Dudley, Kisyński and others, see R. M. Dudley, "On sequential convergence," *Trans. Amer. Math. Soc.* **112**: (1964) 483–507 and J. Kisynski, "Convergence du type L," *Colloq. Math.* **7**: 205–211 (1960). If S is any set, an L^* sequential convergence C on S is a relation between sequences $\{s_n\}$, $n = 1, 2, 3, \ldots$, and members s of S, denoted by $C \lim s_n = s$, such that

i) If $s_n = s$ for all n then $C \lim s_n = s$.

ii) If $C \lim s_n = s$ and if $\{r_m\}$ is a subsequence of $\{s_n\}$ then $C \lim r_n = s$.

iii) If $C \lim s_n = s$ and $C \lim s_n = t$ then $s = t$.

iv) If it is false that $C \lim s_n = s$ then there is a subsequence $\{r_m\}$ of $\{s_n\}$ such that for any subsequence $\{t_p\}$ of $\{r_m\}$, it is false that $C \lim t_p = s$.

Given an L^* sequential convergence C on S, we denote by $T(C)$ the topological space in which a set A is declared open if, whenever $C \lim s_n = s$ and $s \in A$, then $s_n \in A$ for all but finitely many values of n.

Theorem: (Kisyński) *If (S, C) is any L^* space then $C(T(C)) = C$. That is, a sequence $\{s_n\}$ converges to s in $T(C)$ precisely when $C \lim s_n$ s.*

It is easy to see that $T(C)$ is the strongest topology on S with this property. We now focus our attention on B^* with $C \lim s_n = s$ meaning that s_n converges to s in the α topology. Clearly, this is an L^* convergence. We will be done if we can prove that $\gamma = T(C)$. Now γ has the same convergent sequences as α, because by the uniform boundedness principle, an α-convergent sequence must be bounded, and α and γ agree on bounded sets. So γ is weaker than $T(C)$. To prove that $T(C)$ is weaker than γ, let A be a set that is $T(C)$-closed. We must prove that A is γ-closed. But A must be $T(C)$-sequentially-closed and hence γ-sequentially closed. But then A is γ-closed. Note where we have used the fact that $C(T(C)) = C$. Q.E.D.

8. BOUNDED HOLOMORPHIC FUNCTIONS, CONTINUED— DOMINATING AND UNIVERSAL SETS

8.1 Definition: *A subset S of G is called dominating provided that*

$$\sup \{|f(z)|: z \in S\} = \sup \{|f(z)|: z \in G\}$$

for each $f \in B_H(G)$.

8.2 Definition: *Let S be a subset of G. We denote by $M(S)$ the set of all measures that have all their mass in S.*

It is clear that $M(S)$ is a closed subspace of $M(G)$ in the variation norm.

8.3 Definition: *A subset S of G is called universal provided that for each μ in $M(G)$ there exists a measure $v \in M(S)$ such that $v \sim \mu$.*

We say that μ can be swept into S in this case.

8.4 Definition: *A subset S of G is called strongly universal provided that for each μ in $M(G)$ and each $\varepsilon > 0$, there exists a measure v in $M(S)$ such that $v \sim \mu$ and $\|v\| \leq (1 + \varepsilon) \cdot \|\mu\|$.*

8.5 Theorem: *Let S be a subset of G. Then the following assertions are equivalent:*
i) *S is strongly universal*
ii) *S is universal*
iii) *S is dominating.*

We shall see later that in case $G = D$ then the sets S that have these properties are precisely those sets such that almost every point of the unit circle is a non-tangential limit point of elements of S.

Proof: It is trivial that i) $\Rightarrow$ ii). To prove that ii) $\Rightarrow$ iii), suppose the contrary. Then there would exist a point $\zeta \in G \setminus S$ and a function $f \in B_H(G)$ such that

$$f(\zeta) = 1, \quad |f(z)| \leq r < 1 \text{ for } z \in S.$$

Let $v \in M(S)$ be equivalent to the unit point mass ε_ζ at ζ. Then

$$1 = (f(\zeta))^n = \int f^n \, d\varepsilon_\zeta = \int f^n \, dv$$

so that

$$1 \leq r^n \|v\| \qquad \text{for } n = 1, 2, 3, \ldots$$

which is impossible.

To show that iii) $\Rightarrow$ i), which is more difficult, we may suppose without loss of generality that S is a countable set, say $S = \{s_n\}$, $n = 1$, $2, 3, \ldots$, since a dense subset of a dominating set is dominating, and any superset of a strongly universal set is strongly universal.

In this case, $M(S)$ may be identified with the space l^1 of absolutely summable sequences. Let T be the operation of restriction to S of functions $f \in B_H(G)$: $Tf = f|_S$. Since S is dominating, T is an isometric mapping of $H_\infty(G)$ onto a subspace E of the space l^∞ of all bounded sequences. We assert that E is weak-star closed in l^∞ as the dual space of l^1. As in the proof of Theorem 7.5, it is enough to show that E contains all limits of sequences of elements of E. But a weak-star convergent sequence in l^∞ is bounded. If the sequence is $\{Tf_n\}$, it follows that the f_n are uniformly bounded on G. Hence $\{f_n\}$ is a normal family, and passing to a subsequence if necessary, then f_n converge uniformly on compact subsets of G to a bounded analytic function f. But $\lim Tf_n = Tf$ at each point of S, and the assertion is proved.

Now let $N = E^\perp$ be the subspace of l^1 orthogonal to E. Since E is weak-star closed, it is the annihilator of N, that is, $E = N^\perp$. Hence E is the dual space of the quotient Banach space l^1/N. The norm of an element of l^1 as a linear functional on E is equal to its quotient norm.

Now let us choose $\mu \in M(G)$. We may regard μ as a linear functional on $H_\infty(G)$ and consequently on E. As such, it must be continuous with E given the weak-star topology. To see this, we need to prove that the null space of μ, namely, $E_\mu = \{Tf \in E: \int f\, d\mu = 0\}$, is a weak-star closed subspace of E. But a subspace is weak-star closed if it contains all its sequential limits. Let $\{Tf_n\}$ be a weak-star convergent sequence in E_μ with limit Tf. In l^∞, a weak-star convergent sequence is bounded. It follows that $\{f_n\}$ is a bounded sequence, since T is an isometry, and on passing to a subsequence if necessary, we see that $\{f_n\}$ converges boundedly and pointwise to a function in $B_H(G)$ that must be the function f, since it agrees with f on the dominating set S. Since the convergent sequences in $\alpha(G)$ are the bounded and pointwise convergent sequences, we see that $0 = \lim \int f_n\, d\mu = \int f\, d\mu$ so that $f \in E_\mu$, and we have proved that E_μ is closed.

Applying the remarks of the last paragraph but one, we see that μ can be identified with an element σ of l^1/N, and the norm of μ as a linear functional is equal to the quotient norm of σ. In other words, given $\varepsilon > 0$, we can find a measure $\nu \in M(S)$ such that $\nu \sim \mu$ and $\|\nu\| \leq (1 + \varepsilon)\|\mu\|^*$, where $\|\mu\|^*$ denotes the norm of μ as a linear functional on $H_\infty(G)$. But

$$\|\mu\|^* = \sup\left\{\left|\int f\, d\mu\right| : f \in H_\infty(G),\ \|f\|_\infty \leq 1\right\},$$

and by Theorem 7.5, that $H_\infty(G) = (M'(G))^*$, this is the norm of $[\mu]$ in the space $M'(G)$, which completes the proof since $\|[\mu]\| \leq \|\mu\|$.

8.6 Proposition: *There exists a countable dominating subset of G that has no limit point in G.*

Proof: We can write G as the union of a sequence $\{F_n\}$ of open subsets F_n having compact closures, with $F_n \subseteq F_n^- \subseteq F_{n+1}$, $n = 1, 2, 3, \ldots$. Let $C_n = F_n^- \setminus F_{n-1}$, with $F_0 = {}_{\text{def}} \emptyset$, so that each C_n is compact. By the maximum principle, for all $f \in B_H(G)$, we have

$$m_n = m_n(f) = {}_{\text{def}} \max \{|f(z)|: z \in C_n\} = \max \{|f(z)|: z \in F_n\}.$$

so that $m_n \to \|f\|_\infty$ as $n \to \infty$. We let U denote the unit ball of H_∞, and we let $\{\varepsilon_n\}$ be a sequence of positive numbers decreasing to 0. By the uniform equicontinuity of the functions in U, there exists for each n a positive number δ_n such that if $z, w \in C_n$ and $|z - w| < \delta_n$, then $|f(z) - f(w)| < \varepsilon_n$ for all $f \in U$. For each n, we choose a finite subset E_n of C_n such that each point of C_n has distance less than $\delta_n/2$ from some point of E_n. Then

$$\max \{|f(z)|: z \in C_n\} \le \max \{|f(z)|: z \in E_n\} + \varepsilon_n.$$

Hence, $E = \cup E_n$ is the required dominating set with no limit points in G.

8.7 Proposition: *Every dominating subset of G contains a countable dominating subset of G that has no limit point in G.*

Proof: The proof is a continuation of the preceding proof. Let S be a dominating subset of G and let n be a positive integer. For each point z of E_n, we choose a point w of S, if possible, such that $w \in C_n$ and such that the distance of w from z is less than $\delta_n/2$. If there is no such w, we omit this step. Let S' denote the totality of points so chosen for $n = 1, 2, 3, \ldots$. Then S' is at most countable, and has no limit points in G.

To prove that S' is dominating, we fix a function $f \in U$ and choose a sequence $\{\zeta_k\}$ of points in S such that $|f(\zeta_k)| \to \|f\|_\infty$ as $k \to \infty$. We may, by the maximum principle, choose the ζ_k so that they approach the boundary of G. Each point ζ_k is in a set C_n for at least one index n. Let $n(k)$ be the smallest such index. Then $n(k) \to \infty$ as $k \to \infty$ since ζ_k approaches the boundary of G. Hence, if $\varepsilon > 0$ is given, we can choose k so that

$$|f(\zeta_k)| > \|f\|_\infty - \varepsilon, \; \zeta_k \in C_n, \; \varepsilon_n < \varepsilon$$

[where $n = n(k)$. The set E_n contains a point z whose distance from ζ_k is less than $\delta_n/2$. Hence S' must contain a point w whose distance from z is at most $\delta_n/2$. Thus $|w - \zeta_k| < \delta_n$ and hence $|f(w)| > |f(\zeta_k)| - \varepsilon > \|f\|_\infty - 2\varepsilon$. Q.E.D.

It would be good to find geometrical conditions for a set to be dominating. In the case $G = D$, Brown, Shields, and Zeller ["On absolutely convergent exponential sums," *Trans. Amer. Math. Soc.*, **96**: 167–183 (1960)] have found such a condition.

8.8 Theorem: *In order that S be a dominating subset of the open unit disk D, it is necessary and sufficient that almost every boundary point of D be a non-tangential limit of points of S.*

Proof: We suppose first that almost every boundary point is a non-tangential limit of points of S. Let $f \in B_{II}(G)$ be given, and let $\varepsilon > 0$ be given. There is a set of positive measure on ∂D such that $|f^*(e^{i\theta})| \geq \|f\| - \varepsilon$. We use here the fact that a function f in $B_{II}(G)$ has radial boundary values f^* almost everywhere and that the mapping $f \to f^*$ is a one: one isometry of $B_{II}(G)$ into a subspace of $L^\infty(-\pi, \pi)$. At almost all of the points $e^{i\theta}$, $f(z)$ has a nontangential limit equal to $f^*(e^{i\theta})$ and almost all of them are approachable non-tangentially by points of S. Therefore sup $\{|f(s)| : s \in S\} \geq \|f\| - \varepsilon$, and it follows that S is dominating.

For the converse, we proceed by contradiction. First, by Proposition 8.7, we suppose that S is countable, and has no limit points in D. If not almost every boundary point of D is a non-tangential limit of points of S, then there exists a set E of positive measure on ∂D such that no point of E can be approached non-tangentially by points of S. This means that any angle with vertex at a point of E can contain only a finite number of points of S. In particular, this is true for a right angle located with the radius to the point bisecting the angle. So at each point $p = e^{i\theta} \in E$, there is a right triangle Δ_θ with the right angle vertex at p and the other two vertices inside D, having the radius to p as an axis of symmetry, and containing none of the α_n. Thus, there will be a positive number b and a closed subset E_1 of E of positive measure, written $|E_1| > 0$, such that at each point $e^{i\theta}$ of E_1, the altitude of the triangle Δ_θ measured from p has length at least b. Now we let I be a closed arc, whose endpoints are in E_1, such that $|E_1 \cap I| > 0$ and $|I| < b$. Let $G = I \setminus E_1 = I \cap$ comp E_1. Then $G = \cup I_n$ where the I_n are disjoint open arcs of length less than b. We let I_j be one of these arcs with endpoints $e^{i\alpha}$ and $e^{i\beta}$ and draw the two triangles Δ_α and Δ_β. Then the sides of these triangles cross over the interval I_j to form a little "triangle" T_j with one side formed from the arc I_j. If t is a point of I_j, then any of the points of S sufficiently close to t must lie in T_j. We let χ_G denote the characteristic function of G, and we define f by

$$f(z) = \exp\left\{\frac{1}{2\pi} \int_{-\pi}^{\pi} \chi_G(\varphi) \frac{z + e^{i\varphi}}{z - e^{i\varphi}} d\varphi\right\}$$

so that

$$|f(z)| = \exp\left\{-\frac{1}{2\pi}\int_{-\pi}^{\pi}\chi_{1i}(\varphi)\frac{1-|z|^2}{|z-e^{i\varphi}|^2}\,d\varphi\right\},$$

so that $|f(z)| \leq 1$ for $|z| \geq 1$. Similarly for

$$f_j(z) = \exp\left\{\frac{1}{2\pi}\int_{-\pi}^{\pi}\chi_{1i}(\varphi)\frac{z+e^{i\varphi}}{z-e^{i\varphi}}\,d\varphi\right\}.$$

We have $f(z) = \prod f_j(z)$ so that $|f(z)| \leq |f_j(z)|$ for $j = 1, 2, 3, \ldots$.

It is an elementary property of the Poisson integral that

$$\lim|f(re^{i\theta})| = 1 \qquad \text{a.e. } \theta \in E_1. \tag{8.8.1}$$

We may think of $-\log|f(z)|$ as the temperature at z induced by a boundary temperature 1 on G and 0 on the complement of G. We shall now show that

$$|f(z)| \leq e^{-1/2} \text{ for } z \in T_j, j = 1, 2, 3, \ldots \tag{8.8.2}$$

Suppose this is done. We let $t = e^{i\theta}$ be a point of E_1 interior to the arc I at which (8.8.1) holds. Without loss of generality, $t = 1$. Then there is an $\epsilon > 0$ such that if $\text{Re}(s_n) > 1 - \epsilon$, $s_n \in S$, then s_n is in one of the "triangles" T_j. By (8.8.2) $|f(s_n)| < e^{-1/2}$ at all such s_n. Let $g(z) = f(z)e^z$. Then $g \in B_{II}(G)$ and $\|g\|_\infty = e$. But $|g(s_n)| < e^{1/2}$ for $\text{Re}(s_n) > 1 - \epsilon$, and $|g(s_n)| < e^{1-\epsilon}$ for $\text{Re}(s_n) \leq 1 - \epsilon$, so that $\sup|g(s_n)| < \|g\|_\infty$ and it follows that S is not a dominating set. To prove (8.8.2), we need to show that

$$\frac{1}{2\pi}\int_{\alpha}^{\beta}\frac{1-|z|^2}{|z-e^{i\varphi}|^2}\,d\varphi \geq \frac{1}{2} \qquad \text{for } z \in T_j.$$

Now this integral has a simple geometrical interpretation (see R. Nevanlinna, "Eindeutige Analytische Funktionen," pp. 6–7). We extend the line segment from $e^{i\alpha}$ to z until it hits the boundary of D at A and do the same from $e^{i\beta}$ to z to hit the boundary at B. The integral is equal to the arc length from A to B in the counterclockwise direction, divided by 2π. So the minimum of the integral for $z \in T_j$ occurs at the interior vertex of T_j, and this minimum value is

$$\frac{1}{2} + \frac{\beta - \alpha}{2\pi} \geq \frac{1}{2}$$

and our proof is complete.

Following K. Hoffman and H. Rossi ["Weak*-continuous functionals," *Duke J. Math.* (1967)] we give a simple geometric condition that S be a dominating set in D, namely that for some R, the hyperbolic discs of radius R and centers in S cover D. It follows that if S has this property, then almost every boundary point of D is a non-tangential limit of points of S. It would be nice to provide a direct proof of this fact. We use Schwarz's Lemma in the following form: let $d(p, q)$ be the hyperbolic distance on D. If f is an analytic function bounded by 1 in modulus, then

$$|f(p) - f(q)| < \frac{e^{2d(p,q)} - 1}{e^{2d(p,q)} + 1}.$$

We suppose that $\|f\|_\infty = 1$, and we choose a point $p \in D$ so that $|f(p)| > e^R/(e^R + 1)$. There is a point $s \in S$ such that $d(p, s) < R$. Then

$$|f(s)| > |f(p)| - \frac{e^R - 1}{e^R + 1} > \frac{1}{e^R + 1}.$$

If S were not dominating, then there would exist a $g \in H^\infty$ with $\|g\|_\infty = 1$ but $\sup\{|g(s)|: s \in S\} < 1 - n$, where $n > 0$. We choose N so that $(1 - n)^N < 1/(e^R + 1)$ and choose $f = g^N$ to get a contradiction.

8.9 Theorem: *Let $S = \{s_n\}$ be a countable subset of G with no limit points in G. Then S is dominating if and only if there is a measure μ in $M(S)$, $\mu \neq 0$, with $\mu \sim 0$.*

Proof: We suppose first that S is dominating, and we choose $s_0 \in S$ and let $S^* = S \setminus \{s_0\}$. Then S^* is dominating and hence universal. Let $\epsilon_0 = \epsilon_{s_0}$ be the unit point mass at s_0 and let $\nu \sim \epsilon_0$ with $\nu \in M(S^*)$. Finally, we let $\mu = \epsilon_0 - \nu$. It is clear that $\mu \in M(S)$, $\mu \sim 0$, but $\mu \neq 0$. In the other direction, we let μ be a measure in $M(S)$ with $\mu \sim 0$ but $\mu \neq 0$. That is, there exist complex numbers a_n with $0 < \sum |a_n| < \infty$ and $\sum a_n f(s_n) = \int d\mu = 0$ for each $f \in B_H(G)$. We let

$$A(z) = \sum \frac{a_n}{s_n - z} = \int \frac{d\mu(w)}{w - z} \qquad \text{for } z \in G \setminus S.$$

This function is analytic in $G \setminus S$ and has simple poles at each point s_n for which $a_n \neq 0$ since S has no limit point in G. In particular, $A(z)$ is not identically 0, and since G is connected, the zeros of $A(z)$ in G are isolated. We claim that

$$A(z) f(z) = \int \frac{f(w)}{w - z} d\mu(w) \qquad \text{for } z \in G \setminus S, \qquad f \in B_H(G).$$

$$(8.9.1)$$

This is equivalent to the assertion that

$$\int \frac{f(w) - f(z)}{w - z} d\mu(w) = 0 \qquad \text{for } z \in G \setminus S, \qquad f \in B_H(G).$$

But this follows from the hypothesis, since for each $z \in G \backslash S$, the integrand is a bounded analytic function of w. More precisely, the integrand is that function g defined by $g(w) = (f(w) - f(z))/(w - z)$ if $w \neq z$ and $g(z) = f'(z)$. Now suppose that S were not dominating. Then there would exist a function $f \in B_H(G)$ and a point $z_0 \notin S$ such that $f(z_0) = 1$, but $|f(s_n)| \leq r < 1$ for $n = 1, 2, 3, \ldots$. We may further assume that $A(z_0) \neq 0$, since we could otherwise move z_0 slightly, and renormalize f. Applying formula (8.9.1) to f^n, we have

$$A(z_0) = A(z_0)(f(z_0))^n = \int \frac{f(w)^n}{(w - z_0)} \, d\mu(w),$$

which is impossible for large n. Q.E.D.

We remark that the formula (8.9.1) is a discrete Cauchy integral formula for bounded analytic functions. It gives, in effect, an explicit formula for sweeping the unit point measure at the point $z, z \in S$, onto S. It leads to a general balayage formula as follows. If $\rho \in M(G)$ then the measure $\sigma \in M(S)$ given by the following expression is equivalent to ρ:

$$d\sigma(w) = \int \frac{d\rho(z)}{A(z)(w - z)} \, d\mu(w).$$

We shall not give any more details here.

We now show a connection between balayage and removable singularities.

8.10 Definition: *Given a connected open set G' and a compact subset E of G', we let $G = G' \backslash E$. We say that E is a set of removable singularities for bounded analytic functions in G provided that each $f \in B_H(G)$ has a bounded analytic extension $F \in B_H(G')$.*

It is well known (see Ahlfors and Sario, *Riemann Surfaces*, Chapter IV, Section 4 C) that the above property of E is independent of the set G'. Thus, it makes sense simply to speak of E as a set of removable singularities for bounded analytic functions. Such sets E may be considered as thin sets. We shall prove that E is a set of removable singularities for bounded analytic functions if and only if E is so thin that each measure in $M(G)$ can be swept a positive distance away from E.

We begin by introducing the notation

$$E_\epsilon = \{z \in G : \text{distance } (z, E) > \epsilon\} \qquad \text{for } \epsilon > 0.$$

8.11 Definition: *Given G', E, and G as in the preceding definition, we say that a measure $\mu \in M(G)$ is holomorphically free of E if for some $\epsilon > 0$, there is a measure $\nu \sim \mu$ with $\nu \in M(E_\epsilon)$.*

  8.12 Lemma: If every measure μ in $M(G)$ is holomorphically free of E, then for some $\epsilon > 0$, E_ϵ is a dominating subset of G.

  Proof: Let us write E^n in place of $E_{1/n}$. Proceeding by contradiction, if the lemma were false, we could find a sequence $\{f_n\}$ of functions in $B_H(G)$ and a sequence $\{z_n\}$ of points in G such that

$$\sup\{|f_n(z)|: z \in E_n\} \le 4^{-n},$$

$$|f_n(z_n)| > \frac{3}{4}; \qquad \|f_n\|_\infty = 1.$$

We define a measure $\mu \in M(G)$ by

$$\mu = \sum 3^{-k}\epsilon_k$$

where ϵ_k is the unit point mass at z_k. By hypothesis, there is a measure ν that lives in E^N for some positive integer, N, with $\nu \sim \mu$. For $n \ge N$, we have $E^N \subseteq E^n$ and hence

$$\left| \int f_n \, d\nu \right| \le 4^{-n}\|\nu\|. \tag{8.12.1}$$

However,

$$\left| \int f_n \, d\mu \right| = |\sum 3^{-k} f_n(z_k)| \ge \frac{3}{4} 3^{-n} - \left(\sum_{(k<N} + \sum_{k>N)} 3^{-k} f_n(z_k) \right)$$

$$\ge \frac{3}{4} 3^{-n} - \frac{1}{2} 4^{-n} - \frac{1}{2} 3^{-n},$$

which contradicts (8.12.1) for large n.

  8.13 Theorem: Let G' be a bounded open set, let E be a compact subset of G' and let $G = G' \setminus E$. Then E is a set of removable singularities for bounded analytic functions in G if and only if each measure μ in $M(G)$ is holomorphically free of E.

  Proof: It is sufficient to show that E is removable if and only if E_ϵ is dominating for some $\epsilon > 0$. We suppose first that E is not removable, and let $\epsilon > 0$ be given. There exists a nonconstant bounded analytic function f on the complement C of E with respect to the Riemann sphere (See Ahlfors and Sario, "Riemann Surfaces," Chapter IV, Section 4 C). By the maximum principle, we see that

$$\sup\{|f(z)|: z \in G\} = \sup\{|f(z)|: z \in C\}$$

But

$$\sup\{|f(z)|: z \in E_\epsilon\} < \sup\{|f(z)|: z \in C\}$$

since $(E_\epsilon)^-$ is a compact subset of C. Hence E_ϵ is not a dominating set.

Conversely, suppose that E is removable, and choose ϵ with $0 < \epsilon$ $<$ distance $(E, \partial G')$. Then E_ϵ is a dominating set. For, given $f \in B_{II}(G)$, there is a bounded analytic continuation which we still denote by f, into all of G'. But $G' \setminus E_\epsilon$ is a compact subset of G' and thus

$$\sup \{|f(z)|: z \in G'\} = \sup \{|f(z)|: z \in E_\epsilon\}.$$

9. CLOSED IDEALS IN $\beta(G)$

We study here closed ideals in $\beta(G)$ and particularly the case $G = D$, although some of our results have been carried over to some more general regions in the thesis of C. W. Neville, which is now in preparation. We assume familiarity with the theory of bounded analytic functions in D, in particular the fact, which will be explained below, that every bounded analytic function in the unit disc D has a unique representation as the product of an inner function and a bounded outer function [see K. Hoffman, *Banach Space of Analytic Functions*, (1962) Chapter 5]. The radial boundary values of a bounded analytic function f in D, which exist almost everywhere will be denoted by $f(e^{i\theta})$. Inner functions f are characterized by the property that the operation of multiplication by f is an isometry on $H_\infty(D)$, or equivalently that $|f(e^{i\theta})| = 1$ almost everywhere. An inner function f has the representation $f = BS$, where B is a Blaschke product over the zeros of f

$$B(z) = z^p \prod \frac{-\bar{z}_n}{|z_n|} \frac{z - z_n}{1 - z\bar{z}_n},$$

and where S is a singular function

$$S(z) = \exp \left[- \int \frac{e^{i\theta} + z}{e^{i\theta} - z} \, d\mu(\theta) \right],$$

where μ is a non-negative measure that is singular with respect to Lebesgue measure. An outer function Ω has the representation

$$\Omega(z) = \exp \left[- \int \frac{e^{i\theta} + z}{e^{i\theta} - z} \, h(\theta) \, d\theta \right]$$

where $h \in L^1(-\pi, \pi)$. The outer function Ω belongs to $H_\infty(D)$ if and only if ess inf $\{h(\theta): -\pi < \theta \leq \pi\} > -\infty$.

Notation: If $f \in B_{II}(G)$ we denote by (f) the principal ideal generated by f in $B_{II}(G)$, that is $(f) = f B_{II}(G)$.

We shall show that (f) is dense if and only if f has no inner factor. In other words, the topological units in $\beta(D)$ are just the outer functions. And (f) is closed if and only if f is an inner function (multiplied by a unit), as we shall show.

9.1 Theorem: *The principal ideal (f) is dense in $\beta(D)$ if and only if f is an outer function.*

Proof: We suppose first that f is an outer function. It will be enough to show that the constant function 1 belongs to $(f)^-$. We may assume that $\|f\| = 1$. Then

$$f(z) = \exp\left[\int \frac{z + e^{i\theta}}{z - e^{i\theta}} h(\theta)\, d\theta\right] \tag{9.1.1}$$

where $h(\theta) = \log|f(e^{i\theta})|$ is a non-positive integrable function. We let

$$h_n(\theta) = \min(h(\theta), n)$$

$$g_n(\theta) = h(\theta) - h_n(\theta)$$

$$F_n(z) = \exp - \int \frac{z + e^{i\theta}}{z - e^{i\theta}} h_n(\theta)\, d\theta$$

so that

$$f(z)F_n(z) = \exp \int \frac{z + e^{i\theta}}{z - e^{i\theta}} g_n(\theta)\, d\theta. \tag{9.1.2}$$

Now $g_n(\theta) \geq 0$ and g_n converges to 0 in the L^1 metric. Hence

$$\|fF_n\|_\infty \leq 1$$

and $f(z)F_n(z) \to 1$ as $n \to \infty$ for each $z \in D$, and we have proved that $1 \in (f)^-$.

For the converse, we assume that f has a non-trivial inner factor φ. Then $(f) \subseteq (\varphi) \neq \beta(D)$, and our result follows from the next proposition.

9.2 Proposition: *If φ is an inner function, then (φ) is closed in $\beta(D)$.*

Proof: It is enough to show that (φ) is *sequentially* closed. Let us then assume that $\varphi f_n \to g$ in $\beta(D)$. The functions φf_n are uniformly bounded. But $\|\varphi f_n\| = \|f_n\|$ and hence the functions f_n are uniformly bounded. By passing to a subsequence, we may assume that $\{f_n\}$ converges, say $f_n \to f$, in $\beta(D)$. Hence $g = \varphi f$ and it follows that $g \in (\varphi)$, Q.E.D.

9.3 Proposition *If $f \in \beta(D)$ and if φ is the inner factor of f then (f) is dense in (φ).*

Proof: Let g denote the outer factor of f, so that $f = \varphi g$. From the proof of Theorem 9.1, there exists a sequence $\{g_n\}$ of bounded analytic functions such that $g_n g \to 1$ in $\beta(D)$. Q.E.D.

We recall that a function $f \in B_H(G)$ is called a unit if $fg = 1$ for some $g \in B_H(G)$; f is a unit if and only if $|f|$ is bounded away from 0 in G.

9.4 Theorem: *The principal ideal (f) is closed if and only if the outer factor of f is a unit.*

Proof: Let g be the outer factor of f and let φ be the inner factor of f. If g is a unit then (f) is closed, by Proposition 9.2. If g is not a unit, then (f) does not contain φ, for if $\varphi = fh = \varphi gh$, then $gh = 1$. But $\varphi \in (f)^-$ by the preceding theorem, and the proof is complete.

9.5 Theorem: *Every closed ideal in $\beta(D)$ is the principal ideal generated by an inner function.*

Proof: Let I be a closed ideal in $\beta(D)$. If $f \in I$ and if φ is the inner factor of f, then $\varphi \in I$ by Proposition 9.3. It is sufficient to show that I contains one inner function that divides all the other inner functions in I.

Let J denote the closure of I in the Hilbert space H_2 of all functions f analytic in D for which $\|f\|_2 < \infty$, where

$$\|f\|_2 = \lim_{r \to 1^-} \left\{ \frac{1}{2\pi} \int_{-\pi}^{\pi} |f(re^{i\theta})|^2 \, d\theta \right\}^{1/2}.$$

Then J is a closed subspace of H_2 that is invariant under multiplication by z, and so by Beurling's characterization (op. cit. Chapter 7) of the closed invariant subspaces of H_2, J is generated by some one inner function φ_0, that is $J = \varphi_0 H_2$. Consequently, φ_0 divides all the inner functions in I, and it is therefore enough to show that φ_0 itself belongs to I.

Since φ_0 is in the H_2 closure of I, there exists a sequence $\{f_n\}$ of functions $f_n \in I$ such that $f_n \to \varphi_0$ in the H_2 metric. We write $f_n = \varphi_n g_n$ where φ_n is inner and g_n is outer. By passing to a subsequence, we may assume that $\{\varphi_n\}$ converges in $\beta(D)$, say $\varphi_n \to \varphi$. In particular, $\varphi \in I$. Since $\|g_n\|_2 = \|f_n\|_2$, we may use the weak compactness of the unit ball in Hilbert space, and by passing to a subsequence, may assume that $\{g_n\}$ converges weakly in H_2, say $g_n \to g$. In particular $g_n(z) \to g(z)$ for each $z \in D$, since evaluation at points of D is, by the Cauchy integral formula, a continuous linear functional on H_2. It follows that $\varphi g = \varphi_0$. From the next lemma, with $g_0 = 1$, it follows that φ is inner. But from the equation $\varphi g = \varphi_0$ with φ inner and $g \in H_2$, it follows that $g \in H_\infty$, since $H_2 \subseteq H_1$ and the inner-outer factorization of functions in H_1 is unique. Hence $\varphi_0 \in IH_\infty = I$, and we are done.

9.6 Lemma: *Suppose that $\{f_n\}, n = 1, 2, 3, \ldots,$ is a sequence of functions in H_2, that $f_n \to f_0$ in $H_2, f_0 \neq 0$, that $f_n = \varphi_n g_n$ is the inner-outer factorization of $f_n, n = 1, 2, 3, \ldots,$ and that $f_0 = \varphi_0 g_0$ is the inner-outer factorization of f_0. Suppose further that the $\varphi_n \to \varphi$ in $\beta(D)$ and that $g_n \to g$ weakly in H_2. Then φ is an inner function and $g_n \to g$ in the H_2 metric, but g need not be an outer function.*

Proof: It is clear that $|\varphi(z)| \leq 1$ for each $z \in D$, so that $|\varphi(e^{i\theta})| \leq 1$ almost everywhere. To prove that φ is inner, it is enough to prove that $|\varphi(e^{i\theta})| \geq 1$ almost everywhere. If, on the contrary, $|\varphi(e^{i\theta})| < 1$ on a set of positive measure, then for some set of positive measure and for some $\epsilon > 0$, we would have $|\varphi(e^{i\theta})| < 1 - \epsilon$ on that set. It would follow that for each $h \in H_2, h \neq 0, \|\varphi h\|_2 < \|h\|_2$. From the fact that $g_n \to g$ weakly, we see that $\|g\|_2 \leq \lim \inf \|g_n\|_2$. Hence

$$\|g\|_2 \leq \lim \inf \|g_n\|_2 = \lim \|f_n\|_2 = \|\varphi_0 g_0\| = \|g_0\|,$$

$$\|g_0\| = \|\varphi_0 g_0\| = \|\varphi g\| \leq \|g\|.$$

Hence we have equality throughout, so that $\|\varphi g\| = \|g\|$ and φ is consequently an inner function.

Also $\|g\| = \lim \|g_n\|$, and this, plus weak convergence of $\{g_n\}$ to g, implies that $\{g_n\}$ converges to g in norm, since then

$$(g - g_n, g - g_n) \to 0 \quad \text{Q.E.D.}$$

Theorem 9.5 can be used to extend Beurling's theorem on invariant subspaces of H_2 to other spaces of analytic functions, and in particular to the spaces $H_p, 1 \leq p < \infty$. These spaces H_p were treated in this connection by Helson, "Lectures on Invariant Subspaces," 1964. We restrict ourselves to spaces of functions of bounded characteristic, that is, functions that are quotients of two bounded analytic functions. Another characterization of such functions f is given by the criterion,

$$\int_{-\pi}^{\pi} \log^+ |f(re^{i\theta})| \, d\theta \leq m < \infty \qquad \text{for all } r < 1.$$

Every function of bounded characteristic is the product of an outer function with the quotient of two inner functions that have no non-trivial common inner factor. A space of functions is said to be invariant if it is taken into itself by multiplication by each bounded analytic function.

9.7 Theorem: *Let E be a topological vector space whose elements* *are functions analytic in the unit disc D, that satisfies the following five conditions:*

i) *Every function in E has bounded characteristic.*

ii) *If $f \in E$ and $f = \varphi_1 g/\varphi_2$, where g is an outer function and φ_1 and φ_2 are inner functions with no non-trivial common inner factor, then $g/\varphi_2 \in E$.*

iii) *The bounded analytic functions in D form a dense subset of E.*

iv) *For each function f that is bounded and analytic in D, let T_f be the transformation defined by $T_f g = fg$. Then T_f maps E into E and T_f is continuous.*

v) *If S is a closed subspace of E and if $\{f_n\}$ is a uniformly bounded sequence of analytic functions in S that converges pointwise in D to a function f, then $f \in S$.*

Then each closed invariant subspace V of E is generated by an inner function φ, that is, $V = \varphi E$.

Proof: We show first that if f is an outer function in E then the constant function 1 is in the E-closure of $f B_H(D)$. The proof is almost identical with the proof of theorem 9.1. The function f has the representation (9.1.1), where $h(\theta)$ is an integrable function, not necessarily bounded above. We define h_n and F_n as before. Then each F_n belongs to $B_H(D)$ and $f F_n$ has the representation (9.1.2). Hence $\| f F_n \|_\infty \leq 1$ and $f(z) F_n(z) \to 1$ as $n \to \infty$ for each $z \in D$, and therefore by condition (v) of the present theorem, the function 1 belongs to the smallest closed subspace of E that contains $f B_H(D)$.

Now let V be a closed invariant suspace of E, and let $f = \varphi_1 g/\varphi_2$ be the inner-outer representation of f. Then $\varphi_1 \in V$. Indeed, as we have just seen, there is a sequence $\{f_n\}$ of bounded analytic functions f_n such that $g f_n \to 1$ in E. Using (iv), we see that $f(\varphi_2 f_n) = \varphi_1(g f_n) \to \varphi_1$ in E and hence $\varphi_1 \in V$.

We now let $V' = V \cap B_H(D)$ and claim that V' is a closed ideal in $B_H(D)$. First, V' is clearly an ideal. Next, to prove that V' is closed, it is enough to prove that it is sequentially closed. Let $\{f_n\}$ be a sequence of functions in V' such that $f_n \to f$, say, in $\beta(D)$. Using (v) and the fact that $f \in B_H(D)$, we see that $f \in V$.

Now by Theorem 9.5, $V' = \varphi_0 B_H(D)$ for some inner function φ_0. We now prove that $V = \varphi_0 E$. First, to see that $V \subseteq \varphi_0 E$, let f be any function in V, and write $f = \varphi_1 g/\varphi_2$ as its inner-outer representation. From the first part of our proof, we see that $\varphi_1 \in V$, hence $\varphi_1 \in V'$, and consequently $\varphi_1 = \varphi_0 \varphi'$ for some inner function φ' and hence $f = \varphi_0 \varphi' g/\varphi_2$. But from (ii), we have $g/\varphi_2 \in E$, and from (iv) since

$\varphi' \in B_{\Pi}(D)$, we have $g\varphi'/\varphi_2 \in E$ and so $f \in \varphi_0 E$. It remains to show that $\varphi_0 E \subseteq V$, that is, that $\varphi_0 f \in V$ for each $f \in E$. By (iii), there is a net $\{f_\gamma\}$, $\gamma \in \Gamma$ of functions $f_\gamma \in B_{\Pi}(D)$ such that $\{f_\gamma\}$ converges to f in E. But by (iv), $\varphi_0 f_\gamma \to \varphi_0 f$ in E and hence $\varphi_0 f \in V$, since $\varphi_0 f_\gamma \in V'$, for each $\gamma \in \Gamma$, and the result is proved.

9.8 Applications: We now show briefly that the spaces H_p, $1 \leq p < \infty$, satisfy the hypotheses of Theorem 9.7, so that each closed invariant subspace of H_p is generated by an inner function. The case $p = 2$ is, of course the theorem of Beurling we use in the proof of Theorem 9.5, which was used in turn in the proof of Theorem 9.7. The case $p = 1$ was treated by de Leeuw and Rudin ["Extreme points and extremum problems in H_1," *Pacific J. Math.* 8: 467–485 (1958)] and the case of general p was treated by Helson, op. cit. We verify that the conditions (i)–(v) hold.

(i) $\int |f(re^{i\theta})|^p d\theta \leq m < \infty$ for $r < 1$ implies that $\int \log^+ |f(re^{i\theta})| d\theta \leq m' < \infty$ for $r < 1$, so that each $f \in H_p$ has bounded characteristic.

(ii) Passing to the boundary of the disc, we have

$$\int \left| \frac{g(e^{i\theta})}{\varphi_2(e^{i\theta})} \right|^p d\theta = \int \left| \frac{g(e^{i\theta})}{\varphi_2(e^{i\theta})} \varphi_1(e^{i\theta}) \right|^p d\theta$$

since $|\varphi_1(e^{i\theta})| = 1$ almost everywhere.

(iii) The polynomials are bounded analytic functions in D, and by Féjer's theorem, the Césaro means of the partial sums of the Taylor's series of a function in H_p must converge to the function in the metric of H_p.

(iv) It is obvious that T_f maps H_p into H_p and the continuity is also clear since

$$\|T_f g - T_f g'\|_p = \|f(g - g')\|_p \leq \|f\|_\infty \|g - g'\|_p.$$

(v) Let us suppose, by way of way of contradiction, that we have a sequence $\{f_n\}$ of bounded analytic functions S, that $f_n \to f$ in $\beta(D)$, but that $f \notin S$. By the Hahn-Banach theorem, there exists a function $g \in L^q(-\pi, \pi)$ such that $\int f_n(e^{i\theta}) g(e^{i\theta}) d\theta = 0$, but such that $\int fg \, d\theta \neq 0$. But by Hölder's inequality, if we define v by $dv(\theta) = g(\theta) d\theta$, then v is a measure on $(-\pi, \pi)$ that is absolutely continuous with respect to Lebesgue measure. It is not hard to show that this measure v can be swept inside the unit disc, in the sense that there is a measure $\mu \in M(D)$ such that

$$\int f(z) \, d\mu(z) = \int f(e^{i\theta}) \, dv(\theta)$$

for each $f \in B_{II}(D)$. But then for our function f we have $0 = \int f_n \, d\mu$ $\rightarrow \int f \, d\mu \neq 0$, and the assertion is proved by contradiction. Q.E.D.

It is a consequence of Helmer's Theorem that in the algebra of analytic functions in the complex plane, every finitely generated ideal is closed in the topology of uniform convergence on compact subsets. This contrasts with the algebra $\beta(D)$, as the next result shows.

9.9 Proposition: *There is a finitely generated ideal in $\beta(D)$ that is not closed.*

Proof: We outline the proof, which is along familiar lines. By Theorem 9.5, it is enough to construct a finitely generated ideal in $B_{II}(D)$ that is not principal. We choose two sequences $\{z_n\}$ and $\{w_n\}$ of points of D having no points in common, such that $\sum (1 - |z_n|) < \infty$ and such that $|z_n - w_n|$ converges to 0 extremely rapidly as n tends to ∞. Let B_1 be the Blaschke product formed with the zeros $\{z_n\}$ and let B_2 be the Blaschke product formed with the zeros $\{w_n\}$. Suppose that the ideal in $B_{II}(D)$ generated by B_1 and B_2 were principal, with generator f. Then f must have no zeros since B_1 and B_2 have no common zeros. Thus $1/f$ is an analytic function of bounded characteristic, and it follows that for some positive constant c,

$$|f(z)| \geq \exp \left\{ \frac{-c}{1 - |z|} \right\} \qquad \text{for all } z \in D.$$

We would also have $f = g_1 B_1 + g_2 B_2$ for some pair g_1, g_2 of bounded analytic functions. In particular, $|f(z_n)| = |B_2(z_n)| \cdot |g_2(z_n)|$, which is impossible, because g_2 is bounded while B_2 is extremely small at the points $\{z_n\}$.

9.10 Proposition: *There is a maximal ideal in $\beta(D)$ that is not closed.*

Proof: If a maximal ideal is closed, then by Theorem 9.5 it is generated by an inner function φ. But the multiples of φ cannot form a maximal ideal unless φ is a single Blaschke factor. To see this, observe that if φ contains a Blaschke product of at least two factors, or a Blaschke product and a singular function, then the inner function formed by deleting one factor will generate a larger proper ideal. On the other hand, if φ has no zeros, then there is a square root of φ that is an inner function that generates a larger proper ideal. Now let I be the ideal of all functions $f \in B_{II}(D)$ such that $f(x) \rightarrow 0$ as $x \rightarrow 1^-$. There is a maximal ideal J that contains I, and by the above remarks, it follows that J cannot be closed, since the functions in I have no common zeros. Q.E.D.

Now let X be the class of all continuous multiplicative linear functionals on $\beta(G)$, and suppose that G is bounded, so that $e \in B_H(G)$ where $e: G \to \mathscr{C}$ is the identity map, $e(z) = z$ for $z \in G$. If $\chi \in X$, we let χ^* be the complex number $\chi^* = \chi(e)$. Let G^* denote the set consisting of the points of G and of all points that are removable singularities for all functions in $B_H(G)$. We may think of the functions in $B_H(G)$ as being defined on G^*. For $\zeta \in G^*$, let χ_ζ denote the functional of evaluation at ζ, $\chi_\zeta(f) = f(\zeta)$. Then $\chi_\zeta \in X$.

9.11 Theorem: *If G is a bounded region and if ∂G is the union of non-degenerate continua and isolated points, then the only continuous multiplicative linear functionals on $\beta(G)$ are the point evaluations at points of G^*.*

Proof: From the easily proved fact that if G is an open set in the plane, and if p is an isolated point of ∂G then $G \cup \{p\}$ is an open set, we know that if we adjoin the isolated boundary points of G to G, we obtain a region whose boundary is now the union of non-degenerate continua. This new set has no removable singularities in its boundary, for any point $z_0 \in \partial G$ is a non-removable singularity for the bounded analytic function $(z - z_0)^{1/2} (z - z_1)^{1/2}$, where z_1 is a point on the same component of ∂G as z_0, and so this new set is precisely G^*. Now let $\chi \in X$ be given. We first show that $\chi^* = \chi(e)$ belongs to G^*. Indeed, χ^* cannot belong to the exterior of G^*, since then the function f given by $f(z) = (z - \chi^*)^{-1}$ would belong to $B_H(G)$, which would lead to the contradiction

$$1 = \chi(1) = \chi(z - \chi^*)\chi(f) = 0 \cdot \chi(f) = 0.$$

Also, χ^* cannot belong to the boundary of G, for in this case there would be another point $z_0 \neq \chi^*$ in the same component of ∂G as χ^*. Let

$$g_n(z) = (z - z_0) \left(\frac{z - \chi^*}{z - z_0} \right)^{1/n}$$

for some branch of the nth root. Then $g_n \in \beta(G^*)$ and $\chi(g_n) = 0$ since $(\chi(g_n))^n = \chi(g_n^n) = 0$. The functions g_n are uniformly bounded and hence some subsequence converges in $\beta(G)$ to a function that must have the form $c(z - z_0)$, where c is a constant of modulus 1. Hence $c(\chi^* - z_0) = \chi(c(z - z_0)) = 0$, which is a contradiction.

We now know that $\chi^* \in G^*$ and we must show that $\chi(f) = f(\chi^*)$ for all $f \in B_H(G)$. Let us choose $f \in B_H(G)$ and let $g \in B_H(G)$ be defined by

$$g(z) = \frac{f(z) - f(\chi^*)}{z - \chi^*} \qquad \text{for } z \neq \chi^*,$$

with $g(\chi^*) = f'(\chi^*)$. We then have

$$\chi(f) - f(\chi^*) = \chi(z - \chi^*)\,\chi(g) = 0$$

and the proof is done.

The next result is due to Rudin, "Essential boundary points," *Bull. Amer. Math. Soc.* **70**: 321–324 (1964).

9.12 Theorem: *There exists a region G and a β-continuous multiplicative linear functional* χ *on* $\beta(G)$ *that does not come from any point evaluation.*

Sketch of proof: Let $z_n = x_n$ be a sequence of points on the positive real axis that tends to 0 as $n \to \infty$. Let D_n be a small disk with center at z_n and boundary Γ_n, so chosen that the D_n are non-overlapping, and their radii tend extremely rapidly to 0. Let G be the Riemann sphere with the disks D_n removed. Any function $f \in B_n(G)$ has radial boundary values, that we still denote by f on the Γ_n. Define

$$F(z) = f(\infty) + \sum_{n=1}^{\infty} \frac{1}{2\pi i} \int_{\Gamma_n} \frac{f(w)}{w - z}\, dw.$$

Then $F(z) = f(z)$ for $z \in G$. But the series for $F(z)$ converges uniformly and absolutely in the left half-plane $\operatorname{Re} z \leq 0$. We define

$$\chi(f) = F(0)$$

It is clear that χ is a multiplicative linear functional. Also, χ is β-continuous by the Lebesgue dominated convergence theorem. Q.E.D.

ACKNOWLEDGMENT

We express our gratitude to A. L. Shields, B. A. Taylor, and the many others whose work is incorporated into these lectures. There is a great deal of closely related work that we have omitted only because of limitations of space and time.

Author Index

Subject Index